Husserl's Missing Technologies

John D. Caputo, *series editor*

PERSPECTIVES IN
CONTINENTAL
PHILOSOPHY

DON IHDE

Husserl's Missing Technologies

FORDHAM UNIVERSITY PRESS
New York ■ 2016

Fordham University Press has no responsibility for the persistence or
accuracy of URLs for external or third-party Internet websites referred
to in this publication and does not guarantee that any content on such
websites is, or will remain, accurate or appropriate.

Fordham University Press also publishes its books in a variety of
electronic formats. Some content that appears in print may not be
available in electronic books.

Visit us online at www.fordhampress.com.

Library of Congress Cataloging-in-Publication Data
Names: Ihde, Don, 1934– author.
Title: Husserl's missing technologies / Don Ihde.
Description: First edition. | New York, NY : Fordham University Press,
    2016. | Series: Perspectives in Continental philosophy | Includes
    bibliographical references and index.
Identifiers: LCCN 2015034491| ISBN 9780823269600
    (cloth : alk. paper) | ISBN 9780823269617 (pbk. : alk. paper)
Subjects: LCSH: Husserl, Edmund, 1859–1938. | Technology—
    Philosophy. | Phenomenology.
Classification: LCC B3279.H49 I358 2016 | DDC 193—dc23
LC record available at http://lccn.loc.gov/2015034491

Printed in the United States of America

18 17 16   5 4 3 2 1

First edition

In memory of my brother, Jon, and a celebration of the co-invented bricolage technologies we forged on the farm long ago

Contents

Preface
First Encounters with Husserl's Phenomenology

In 2011, the Society for Phenomenology and Existential Philosophy celebrated its fiftieth anniversary with panels and symposia giving retrospectives on its history. Indeed, with this organization, phenomenology was marking its institutional beginnings, from 1962, as a distinctly minoritarian movement in North American philosophy. Most departments were dominantly analytic departments, and that was particularly the case with graduate departments.

I, along with many, many of today's recognizable American phenomenologists, was a graduate student then. I missed the first meeting of SPEP but got to the second at Northwestern University when it decided on its name. The business meeting was long and contentious, with the main issue one of which name would take precedence—phenomenology or existential philosophy. Phenomenology won. But since I was one of the relatively early post–World War II generations of graduate philosophers, I like many others came to phenomenology in a reverse way. For many undergraduates the postwar rage was "existentialism," and we were reading Camus, Nietzsche, Kierkegaard, and Sartre, and some read Heidegger and Merleau-Ponty. Husserl was basically unknown. I was, with my peers in the '50s, reading existentialism, and this continued through theological school. Doing an MDiv at Andover Newton, which was part of a Boston consortium of theological schools, I kept on reading Kierkegaard and Pascal, and I added Berdyaev, who became my thesis figure. I also spent quite a few courses with Paul Tillich at Harvard Divinity School. It was through Tillich that I was introduced to Heidegger. But still no Husserl. But 1962 was a crucial year—not only was it SPEP's inaugural, but Herbert Spiegelberg published his two-volume work, *The Phenomenological Movement: A Historical Introduction* (published in 1960, but I got my copy 1962). By then I was searching for a dissertation topic, and Spiegelberg was invaluable. He made it apparent that existentialism was itself dependent on phenomenology, and phenomenology in its classical form was invented by Edmund Husserl. I began to read Husserl, although I picked Paul Ricoeur as my dissertation figure. At this same time, Erazim Kohak, new from Yale, had begun to teach Husserl at Boston University, and I took those classes.

Here I must throw in a sort of romantic anecdote concerning Husserl. In 1961 I had purchased fifty-six acres of forest land in Weston, Vermont. I built a small, three-room log cabin—one room at a time; it was just about the size of Heidegger's hut—and too far from electricity, so I spent summers, 1962 on, reading both Heidegger and Husserl by kerosene lamplight. That didn't seem too strange to me since I had grown up on a Kansas farm with kerosene lighting. The farm got electricity only after World War II! I was hooked on phenomenology, and the courses I taught back then, first at Boston University as a graduate lecturer, then post-PhD at Southern Illinois University, included *Being and Time,* and *Cartesian Meditations* and *Ideas I.* (In passing, note that I used Hannah Arendt's *The Human Condition* and Herbert Marcuse's *One-Dimensional Man* in other courses.)

One thing began to be clear to me regarding Heidegger's and Husserl's different narrative and linguistic strategies. The common complaint about Heidegger was that he was obscure, difficult to understand, and etymologically eccentric. The complaints about Husserl were that he was an idealist, subjectivist, and a kind of antirealist. Yet as most professors learn, the deepest way to gain deeper understandings of a philosopher is to teach them. What I discovered was this: With Heidegger, yes, he was initially obscure and hard to penetrate, but he was creating his own linguistic world. And once you began to enter that linguistic world, the clearer it became. Indeed, a common flaw of Heidegger followers is that they begin to speak like him! And as I much later came to realize, once "inside" it becomes difficult for "true believers" to have distance. With Husserl the strategy was quite different. Husserl often used carpentry examples—and he was indeed a philosophical builder. His scaffolding consisted of his complicated series of *reductions: epoche, eidetic, transcendental.* But his linguistic strategy was one that began with the vocabulary of precisely the philosophy he wished to deconstruct. Thus with Descartes, Husserl speaks of *ego cogito,* of things, extensions, shapes, and the like, and although his aim is to overcome, indeed invert in many cases, the Cartesian meanings, the vestigial origin language carries its own momentum. This Cartesian tendency remained with him, as I show in this work. I note this phenomenon here with respect to technologies and instruments.

So my first encounter with Husserl and his version of phenomenology goes back to the early to mid-sixties. Within a decade after graduate school, I began to discover the philosophical challenge of technologies. Here is how it began: In my first year (1964) at Southern Illinois University, I was assigned to teach in a team-taught, interdisciplinary honors course. The theme for that year was "leisure in a work society." Diverse readings posed the thesis that as society became more and more technologically advanced, leisure time would grow, and some writers were utopian enough to hold that such advanced technological societies could become a sort of "second Greece" with leisure time to become highly creative. Remember, I was also reading and teaching Arendt's *The Human Condition*, and this utopian twist clearly did not fit her take on technology and society. Rather, she and her first husband, Günther Anders, were very much under the sway of Heidegger, who clearly was not utopian! It was at this very juncture that I made my first Husserlian move.

Husserl's advice was to *do phenomenology*! So, I set out to do what I then called a "phenomenology of work." What do we do when we are actively engaged with some project? What emerged was a growing recognition that from the most ordinary and even trivial activity, we engage with technologies. I asked students to keep a sort of action-technology diary and count up and itemize our encounters with technologies. These were to be descriptively analyzed in phenomenological fashion. The result was overwhelming. Typically, the first moment of awakening, becoming conscious, was linked to an alarm clock. On to the plumbing system and the first flush of the day, and first washing of face and hands or full shower. Breakfast engaged the toaster, cutlery, refrigerator, lighting system, and so on. The classroom—then, was ballpoint pen and notebook—today, laptop or iPad. And on, and on, for literally hundreds of human-technology encounters. This exercise grew into my early attempts to phenomenologically account for a variety of human-technology relations. From these early attempts to do phenomenology, however, there came a big surprise. Relations to, with, and through technologies *in use* turned out from the beginning to be nothing like "Cartesian" relations to things or objects. The Cartesian analysis was one of dealing with *res extensa,* things as having predicates (colors, weights, sizes, shapes, and the like) and as external and "out there." The alarm clock, awakening me with its noise, did not get experienced as an out-there object; it did not first appear as an object with a dial, an instructive call to my slumbering self. Although below an animate "other," its machinic call was quasi other and auditory.

Even more were tools and instruments not primarily "objects," but they were means by which one could interact with an environment or world. Eyeglasses "withdraw" as objects; and through hammers I experience the nail being driven into the wood. Now in one Husserlian sense, the reductions should "bracket" the natural attitude (which is a kind of Cartesian metaphysics in ascription), and thus there is no phenomenological call for a tool to be first or primarily an object. Thus as my analyses matured into the *phenomenology of technics*, or experiential-actional relations to, through, and with technologies, the role of these within intentionality became more and more praxical, less and less inert or objectlike. But although there are probably only a finite number of human-technology relations, each technology has its distinctive set of affordances and constraints. And that is one proper task for a philosophy of technology to undertake.

So now in later life, I am looking back at my philosophical tutors. I am, as it were, paying my philosophical debts. I did this first with *Heidegger's Technologies* in 2010, and I am now doing the same with Husserl, and I am here doing a reflection on how he dealt with such a subject matter. Both

Heidegger and Husserl I regard as inspirations and fore-fathers, but after many years of work in the philosophy of technology, I also feel the duty for a critical reappraisal.

For this book I have chosen to do endnotes by chapters. For those quotations from Husserliana, the *Briefwechsel,* and the *Nachlass,* for footnotes I have adapted the accepted abbreviation style of HUA, volume number, year, page, and so on. The full titles of volumes are given in a headnote to the Notes section. My three translators, Betsy Behnke, Frances Bottenberg, and Søren Riis, are indicated in the notes by their initials.

I would never claim to be a deep, philological scholar. Although Husserl remains one of my most profound philosophical influences, my own style of philosophizing is problem-oriented. I do not spend the hours of time needed for philological work in archives, libraries, and the like. I appreciate those who do this and have here frequently called on them (see Acknowledgments at the end of the book). But I do take my "godfathers" seriously and in Husserl's case found his variational theory to be profound and rigorous—even if it leads, as it does here, to the deep criticism that he often failed to take his own advice.

Before I leave this preface I would like to comment on my writing style. My stylistic hero has always been Søren Kierkegaard, who anticipated the postmodern love for irony and its style of humor. His images resonate—like the fellow who puts a ball in his floppy back pocket, which slaps him on the thigh as he walks to remind him that "I'm not crazy, I'm not crazy." But I have also learned from contemporary feminists. Some have criticized the impersonal, third person style of expression that came to characterize so much modern scientific and philosophical writing. Donna Haraway—a friend for over thirty years—has honed a descriptive style I find enticing. And Susan Bordo, a former student and alumna of Stony Brook, has done more to personalize her writing, to a degree that I myself cannot successfully follow. Her reflections, titled "My Father, the Feminist," who by contrarian example stimulated her, also echo Kierkegaardian irony. My own style, which often reverts to first person voice and includes personal anecdotes, nevertheless leads to further conclusions, as Vivian Sobchack has pointed out in the section titled "Ihde's Unique Voice," in her essay "Simple Grounds: At Home in Experience," in Evan Selinger's festschrift, *Postphenomenology: A Critical Companion to Ihde*. Philosophically there is an implicit point to be made: If human knowledge is situated, embodied, and perspectival, as both classical and postphenomenology hold, then I hold that personalization, at least to some degree, should be acceptable. Of course, it doesn't end there. I, and many others, learned from Herbert Spiegelberg's intersubjective approach to phenomenology in his famous Washington University workshops. And with the turn to science-technology studies one learns through these self-critical and often ingenious experimental practices to respect that style of knowledge as well.

I close with what is a telling anecdote: Writing this book is taking place in 2014, which turns out to be a bumper year for science-technology movies. *Gravity* and *Interstellar* were both science fiction adventure films, the first about a disaster aboard a space station, the second a search for a habitable planet to replace ours. Then there was *The Theory of Everything* and *The Imitation Game*, both biographical movies about famous men: Stephen Hawking in the first, Alan Turing in the latter. Many of the reviews and responses, especially in science-oriented sources such as *Science, Scientific American,* and the *New York Times* science section, expressed worries about how accurate the "science" was. There was even a book about the science of *Interstellar*. But in no source have I seen any worry about whether science's social history was accurately depicted! My guess is that the science establishment still likes a mythology that resides in an outdated fictive social history. This is the myth of the "great man, great mind, individually discovering or inventing something great." Of course, filmmakers like this simplistic myth as well. Although both Hawking and Turing are so depicted in the films, the Hawking portrayal was particularly egregious. Hélène Mialet has a brilliant book, *Hawking Incorporated* (2012), which does an ethnographic—and I would say also postphenomenological—analysis of Hawking's "three bodies": his physical body, badly disabled by ALS disease; his technologically mediated body of voice synthesizer and medical apparatus-loaded wheelchair; and his socially distributed body, which is the large network—from nurse to ghostwriters and travel arrangers—which allows him to be the public giant he is. As she shows, every year Cambridge University would assign four of the brightest and best research assistants to work with him. She describes how they learn to decipher his gestural language of eye, lip, and head motions; they do much of his writing; and above all, since he himself admits he cannot do the calculations called for in astrophysics, they calculate for him. And interestingly, despite the well-known multiple-author convention in scientific publishing, the articles that appear do so under Hawking's name alone (another Kierkegaardian irony?). Although this anecdote may appear odd here, it is actually telling. Science, technology, and technoscience are filled with sociohistorical myths, and sometimes these pervade philosophy as well—as I think once dominant Cartesianism partially pervades Husserl.

Husserl's Missing Technologies

Introduction
Philosophy of Technology, Technoscience, and Husserl

This book is about and in response to Edmund Husserl's phenomenology and his notion of science. No one would claim that Husserl was a founder of philosophy of technology, although his writings about science are substantial and persistent. Readers may note that my *Heidegger's Technologies: Postphenomenological Perspectives* (Fordham, 2010) was a response to Heidegger's foundational work in precisely philosophy of technology. This book parallels that work by turning to Husserl on technologies despite his ambivalence and lack of focal attention to technologies or even to instruments in science. Husserl remained throughout his life a believer in the primacy of science with regard to technology. In that respect he remained modernist.

In my own case, I have now written about both technology and science—*technoscience*—for forty years. And I acknowledge my deep debt to both Heidegger and Husserl with their work in creating a hermeneutic and phenomenological style of analysis that, despite opposition by the dominant traditions of twentieth-century philosophy of science, holds, I believe, a deep set of implications for the understanding of contemporary technoscience. The scene here is complex, precisely because in the sciences, in the philosophies of science and technology, there have been many twists and upheavals just within the last century to the present alone. Years ago, in fact before I began work on my PhD, I studied theology with Paul Tillich, whose notion of *heteronomous time* has reemerged to stimulate my recent thinking. Temporal heteronomy is a sort of time-layering which in certain instances can lead to internal conflicts within thought and practice such that consistency and uniformity is impossible. I find this notion helpful with respect to Husserl. There will be several variations on this theme. One will be to recognize Husserl in his own setting and subsequently reread him retrospectively. A second will be to parallel his and our times. His work, taking significant shape early in the twentieth century, took place in an intellectual milieu which was still thoroughly modernist but which had reached limits not yet visible until much later in the century. Indeed, much of Husserl's work could be said to be an attempt to overcome and transcend early modern epistemology.

That epistemology was first forged at the same time as the emergence of early modern science, primarily in the seventeenth century. Within science a key figure was clearly Galileo Galilei. Galileo in the early seventeenth century was the discoverer of new phenomena via the telescope that served to convert him from Ptolemy's geocentric to Copernicus's heliocentric systems. This conversion, particularly for purposes here, was made possible by then new instrumentation, optical technologies. Telescopes and microscopes, both constructed by Galileo himself, were the mediational means of transforming and enhancing vision itself. In this sense technologies were a means of discovery, of bodily-perceptual discovery. And Galileo had a great deal to say about telescopic vision, and he concluded that it was superior to bodily or "eyeball" vision. I will here do two in-depth analyses of Husserl in relation to Galileo with the focal technology of the telescope as my fulcrum for bringing together the philosophical themes and issues that surround embodied vision. Husserl used Galileo as his figure for the emergence of early modern science. I shall look closely at Husserl's Galileo.

The early modern philosophical figure who plays an equally central role for Husserl is René Descartes, who may be said to have been a primary inventor of early modern epistemology. This epistemology is explicated, quite deliberately by both Descartes and John Locke, through the metaphor of the camera obscura, which was taken as a model for both eye and mind (see the appendix at the end of this book). There is a kind of irony here: The camera, one of the optical technologies that played an important role in the Renaissance in art, later adapted to science, is also an optical technology, used to model scientific rationality itself. Both early modern science and early modern epistemology are permeated with optics, both literally and metaphorically. Thus a technology is at least indirectly relevant for early modernity. Interestingly, although Descartes did not forefront technology in his concept of a geometrical science, he was highly appreciative of science's instruments. Contrarily, Locke remained skeptical of instruments for the production of scientific knowledge. For Husserl, both Galileo and Descartes play central roles in his attempt to deconstruct early modern epistemology. But their technologies do not fall under his gaze. From the *Cartesian Meditations* to the *Crisis* Descartes and Galileo haunt Husserl's attempts to invert modern epistemology. Phenomenology was to be Husserl's contrarian epistemology, which, in turn, was to be a new rigorous science in its own right. Phenomenology was to be one of the early twentieth century's new philosophies struggling to overcome early modernity.

Husserl wrote about Galileo and Descartes. He did not write about John Dewey. Pragmatism, however, was phenomenology's American counterpart as a new philosophy determined to undermine early modern epistemology. And it later became adapted to a style of analytic neopragmatism, as Richard Rorty has so well shown. Husserl was, in fact, born in the same year as John Dewey, 1859. Phenomenology and pragmatism were thus at least chronological twins through their founders, and I contend, both were contrarian with respect to early modern epistemology, although Dewey and Husserl took different paths in this endeavor. In this book I shall be looking first at Husserl in relation to his chosen adversary-predecessors, primarily Galileo but also Descartes. But although Husserl was aware of pragmatism in its formation by William James, it is not clear that Dewey played any significant role in his thought. So I shall in the second half of this book, turn to the Deweyan-inspired form of pragmatism to see if a productive dialectic can be enticed. For almost two decades I have argued that pragmatism and phenomenology can mutually influence each other. What I call *postphenomenology* can serve to better analyze contemporary technoscience. In this context I show how Dewey and Husserl can be complementary. What is clear is that both pragmatism and phenomenology made *experience* central. But Dewey and Husserl had very different notions about what counted as experience. I shall look at both—Husserl with his *consciousness-centered* version of experience and Dewey with his *experimental* and *instrumental* version of experience. Both, I hold, attacked early modern epistemology and substituted an *interrelational ontology* for the subject-object model of modernity.

Antimodernities

The twentieth century was marked by a series of "antimodernities," including postmodernism, a family of French-inspired trends including poststructuralism, deconstruction, and its kin. There were also attacks on Enlightenment humanism, including Heidegger's, and later including posthumanism and transhumanism, and all these remain alive in this century. However, I find two framings of modernism most relevant to the early antimodernism of phenomenology and pragmatism: one comes from Bruno Latour, *We Were Never Modern* (1993) and the other from Paul Forman, *History and Technology* (2007). These two thinkers are highly relevant to my thesis here precisely because both are doing technoscience studies.

My reason for turning to Latour and Forman is that both relate modernism to science and technology. The antimodernisms I refer to above tend to concentrate on humanities, artistic, and religious concerns, presumed "deaths" or endings of master narratives, Enlightenment humanism, the subject and author, the linguistic turn, are all most often discussed in humanities and arts contexts. Indeed, when postmodernists, feminists, and the new social sciences turn to science, the result was usually deep criticism from the promodern *science community,* the so-called science warriors. For example, Paul Gross and Norman Levitt, drawing from a conference, published *Higher Superstition: The Academic Left and Its Quarrels with Science* (1994) in which they identified feminists, postmodernists, and literary critical theorists as enemies of science. Later, the "science wars" surrounding Alan Sokal's hoax in *Social Text* (1996) continued to describe humanists as scientifically naïve, as relativists, and irrational. The so-called science warriors were also staunch defenders of modernism. I note in passing that my *Postphenomenology: Essays in the Postmodern Context* was published in 1993.

Bruno Latour, who with Steve Woolgar, launched a new sociology of science approach with *Laboratory Life* (1979), depicts how the sciences relate to modernity. Latour went farther in *We Were Never Modern* (1993). Latour argues that science operates within what he calls "The Modern Constitution," in which "modernity is often defined in terms of humanism. . . . This habit itself is modern, because it is asymmetrical. It overlooks the simultaneous birth of 'nonhumanity'—things, or objects, or beasts."[1] This entails a dichotomy that he describes as a work of *purification*. On one side there is Nature (all the nonhumans) and on the other Humans (culture or society).[2] Although this echoes Lévi-Strauss's earlier nature/cultural distinctions in structuralist anthropology, it is also part and parcel of early modern *metaphysics*. And for the most part it is identifiable as the metaphysical framework of the so-called science defenders of the midcentury "science wars."

I shall not here go on to deal with Latour's solution, but he does identify a problem that is relevant to Husserl's missing technologies. According to his modern constitution, technologies are obviously enigmatic: They are objects, things, nonhumans, which *do not fit well with the purification of modernist metaphysics!* Technologies—more particularly traditional and preindustrial technologies—are Nature insofar as their materials are natural products. The hammer has a wooden handle and its iron head is manufactured from natural ores; the telescope's tube is likewise metal and its lenses glass. But except for found uses, such as finding a rock to use as a pounder, technologies are also manufactured, deliberately made by humans, and thus are also cultural. For Latour, beasts are also part of Nature, nonhuman. But of course beasts also make and use technologies. Chimpanzees have various styles of pounder-anvil technologies for cracking nuts, although the amount of manufacturing going into these is minimal. Animals do shape tools, but the shaping remains relatively simple and rarely compound. Humans, from Stone Age tools such as the bifacial Acheulean hand axes (see next chapter) are highly shaped. And spears with stone tips have sinew attachments to polished shafts. And such hybrids proliferate into modernity. Indeed, since industrial times even the base materials can be fabricated— plastics are not found in Nature. 'The drastic conclusion that lurks in the "Modern Constitution," as Latour calls it, is that it has no sufficient way of dealing with technologies! Latour also notes that modernity is a break from the past, premodernity. Such a break was in fact a major point of contention in early philosophy of technology, producing very strange bedfellows. For example, both Martin Heidegger, clearly a "continental" philosopher, and Mario Bunge, an early analytic philosopher of technology, have argued that modern technology is radically different from premodern or traditional technology. Both agree that premodern technology is a *craft* type technology, whereas modern technology is industrial and scientific. For Heidegger this distinction reveals much of his romanticism for craft technologies, which he prefers to modern technologies, whereas for Bunge modern technologies are simply "applied science."

Here, then, I turn to Paul Forman, a historian of science and technology and a curator of the Smithsonian. In a special issue of *History and Technology* (2007), he announced the following thesis: In modernity, it has always been presumed that science was primary. Indeed technologies were presumed to be part of science, but with postmodernity there was a reversal by which technology becomes forefronted and primary—here is what he says:

> The abrupt reversal of culturally ascribed primacy in the science-technology-relationship—namely from the

primacy of science relative to technology prior to circa 1980, to the primacy of technology relative to science since about that date—is proposed as a demarcation of postmodernity from modernity; modernity is when "science" could and often did denote technology too; postmodernity is when science is subsumed under technology.[3]

Forman goes on to deal with a number of figures, showing how each is situated on one or the other side of the modernity/postmodernity divide. By his count, Husserl (and Dewey) remain modernists since both privilege science over technology. Heidegger would be on the postmodernist side—but note he well precedes Forman's 1980 date!

With Latour and Forman, both of whom introduce the issue of modernity with specific reference to science and technology, we reach a turning point. Both see modernity as a problem, particularly for science-technology analysis. In Latour's case the hybrid nature of technology is impossible to incorporate into the implicit metaphysics of the modern constitution. It is clear that metaphysically technology should be part of the nonhuman or artifactual realm. But in use—to which I will turn—it is, in Latour's term, an *actant* or better, an interactant. And in his case it becomes one side of an "actor network," which I will not pursue farther here.

In Forman's case, technology, or at least science's technologies, are subsumed within science. The implication allows technologies to be both artifactual and natural, but not interrelational. So, with both Latour and Forman, modernity blocks a thematic analysis of technologies. And it is here that both phenomenology and pragmatism become relevant. Both develop what I shall call an *interrelational ontology,* which alternatively is a praxis-oriented analysis based on understandings of human experience in relation with, to, and through technologies. Both also turn to a praxis analysis of technologies in use in human experience.

One last preliminary step needs to be taken here—a brief schematic look at Husserl's technologies. Here I limit myself to what I think all Husserl scholars would agree upon. In his published lifetime work, references to technologies are sparse. Indeed, as many readers of Husserl note, he is not given to detailed illustrations or examples of any kind with any frequency, and many examples he uses are mundane. Here I am particularly grateful to a fellow Husserl reader, Betsy Behnke, who provided me a most thorough set of references to Husserl's technologies. She gave me several references from the published works (*Ideas II, Preface to the Logical Investigations*), but these are scarce. And as will be seen in Chapter 2, I had already used references from *The Origin of Geometry* and the *Crisis*. But, as Behnke points out, rather more numerous mentions occur in the *Briefwechsel*. These are informal references to the "technological milieu the Husserl family lived in . . . references to telegrams . . . vehicles . . . [and] Malvine thanks her children for . . . sending photographs of the grandchildren. But these are not philosophical discussions of such matters."[4]

I shall return to these references again, but while mundane, the informal references do show an attitude, particularly with respect to the *modern technologies* being referenced. Husserl is clearly accepting, sometimes even praising, improvements. He is clearly "modernist" in contrast to the suspicious "antimodernism" of Heidegger.

I now return to the structure and organization of this book. As I have noted in the preface, the Husserl Circle is the primary organization in North America for the serious study of Edmund Husserl. It was founded in 1969. I am unsure what year I first attended the circle, but it was early, and I have frequently participated since. Indeed, the core chapters for *Husserl's Missing Technologies* were, in several cases, as noted, first presentations at the Husserl Circle. This book collects some of those presentations that are relevant to Husserl on science and technology, a central concern of mine since the early 1970s. My first book on the philosophy of technology, *Technics and Praxis: A Philosophy of Technology,* was published in 1979, and one of its primary theses concerned issues surrounding science's technologies—instruments. I argued then that science is materially embodied in its technologies and instrument uses. In Forman's framework this places me in the postmodern world with respect to science-technology. This was the era in which what today is termed the "practice turn" and the "material turn" was taking place, although I drew my resources primarily from phenomenology, hermeneutics, and pragmatism rather than the social sciences. And it was from this perspective that I will here reread Husserl.

This book begins with two chapters that look directly at what Husserl had to say about science's instruments and their relation to science practice. Three chapters were originally presented in shorter versions at Husserl Circle meetings. One chapter each were also conference papers. The first chapter looks at what Husserl had to say about telescopes and microscopes and refers to his *Nachlass* now published in the 2007 Husserliana volume 39. That chapter was presented at the Husserl Circle at Dartmouth College in 2012. The second chapter, "Husserl's Galileo Needed a Telescope!", turns to the later Husserl with his turn to praxis-originated thinking and the technology of writing from his late works. That chapter originally was presented at the Husserl Circle in Lima, Peru, in 2002. The third chapter looks at embodiment skills and the reading-writing technologies of Husserl's time. This allows me to show some of the rationale for an embodiment postphenomenology (and provides a parallel to my previous work on the same subject in *Heidegger's Technologies*). The fourth chapter returns to a paper, "Whole Earth Measurements: How Many Phenomenologists Does It Take to Detect a Greenhouse Effect?" This was first a presentation at the Society for Philosophy and Technology in Puebla, Mexico, in 1996 and was subsequently published and reprinted in a number of publications. Here I revisit the themes raised but in the specific context of modernity and postmodernity noted above. It has been thoroughly rewritten.

During this same period, my interests had also turned to pragmatism, and in particular John Dewey and his influence on the neopragmatism of contemporary analytic philosophy, particularly as interpreted by Richard Rorty. Chapter 5, "Dewey and Husserl: Consciousness Revisited," is my direct attempt to compare Dewey and Husserl. Here I view the contemporary revival of interest in consciousness from the equally contemporary emergence of animal studies as relevant to the usual contrast between phenomenology and pragmatism and one between transcendental and naturalization trajectories in philosophy. This chapter was first presented at the Husserl Circle at Charles University, Prague, in 2007.

Continuing the pragmatism-phenomenology concerns, Chapter 6, "Adding Pragmatism to Phenomenology," makes a more contemporary move. Here I look at the contemporary phenomenology-pragmatism situation concerning technology. This chapter was presented at a pragmatism and philosophy of technology conference organized by Junichi Murata at the University of Tokyo in 2003. I conclude the phenomenology-pragmatism section with Chapter 7, "From Phenomenology to Postphenomenology." I have finally added as an appendix, my Millennium Essay from *Nature* magazine, 2000, on epistemology engines, which shows how early modern epistemology based itself on the model of the camera obscura.

I now return to Tillich's *heteronomy*, which I mentioned previously, as it plays a role in the multiple layers of interpretation I undertake in this project. Husserl obviously belongs to his own time, especially with respect to his take on modern science. He, and Heidegger as well, took science to be a *mathematizing project*. It was out of science-as-mathematization that much early twentieth-century philosophy of science characterized science primarily as a *theory*-biased project with *propositional* truth as its product. Only later, by late mid-twentieth century, did the *praxis* and *materiality* perspective on science emerge. Yet that is my time, and I belong to this time. So I take this layer of time and reread Husserl retrospectively through this perspective. I do this double layering in several ways, Thus this double-layered historical time appears in several chapters. Then, too, the role of phenomenology—I claim—must itself adapt to such time layers and changes, and that is where *post-phenomenology* enters the interpretation. Adapting classical phenomenology to a praxis and materiality perspective calls for different emphases. I trace these out particularly in the later chapters.

Where Are Husserl's Technologies?

As I begin the analysis of Husserl's technologies, I now raise the stakes. It was noted previously that his references to technologies, by wide consensus of Husserl scholars, are sparse, and one can add, for the most part, these are mentioned in passing without serious or in-depth philosophical analysis. Thus I am calling these his "missing technologies."

In what sense are technologies missing from Husserl? In a first, very ordinary sense, they are mostly missing from his writings. He infrequently refers to what today we would call technologies, and in this sense one has to conclude that they do not take up much reflective interest—indeed, much philosophical interest. This, in itself, is not unusual, since this would be true of most philosophers, and most philosophers of his time. And perhaps even more specifically in Husserl's case, because his interests were those from the more abstract cognate disciplines—logic, mathematics, psychology, idealization, and consciousness.

And, if one asks, with what technologies would Husserl necessarily be familiar, the sense of minimalism would have to be raised even more. I do want to point out that Husserl was clearly very familiar with one optical technology—eyeglasses. Most photographs of Husserl show him wearing the old circular style of academic eyeglasses And, of course, he was familiar with the tool technologies used for reading and writing. And, he was clearly familiar with typewriters and the carbon copies used in conveying copies to colleagues and friends. He himself used a pen; he died well before word processing, and although he showed no antipathy to typewriters (whereas Heidegger did), he simply followed the then dominant practice of using a pen. His practice produced the some 40,000 pages of German shorthand notes, which still today are being transcribed and published. I shall be referring to the recently published *Nachlass* manuscripts in this chapter. In Chapter 3 I return to the reading-writing technologies and Husserl's problems with these. Husserl seems to simply take these material mediating technologies for granted. Herbert Spiegelberg, that foremost historian of phenomenology, once told me that Husserl, writing standing up, would be interrupted midmorning by his wife, who would bring in coffee and cookies, and Husserl would regularly comment, "Ach, was gibt? Kuchen, Ja." As it turns out, even this standing practice is debatable, and I will return to it again.

This indicates that it is unlikely that Husserl gave attention, whether ordinary or phenomenological, to ordinary use-technologies. So, in these ordinary senses Husserl is missing his technologies. In this chapter I will be giving special attention, not to such ordinary technologies, but to special technologies used in science practice, to instruments. Telescopes and microscopes, optical technologies, were of high importance in the early formation of science, and as the recently published *Nachlass* shows, Husserl did have things to say about these instruments that related to the "near-far world." But allow this to be framed by taking note of some historical comments on technologies.

Technologies

One aspect of technologies not often discussed is that they become obsolete, are abandoned, and forgotten—at different speeds. I will begin with variations on these contingencies:

> The Acheulean hand axe may be the tool longest used by humans. Named for its first example found by the Acheul River in France, 1859, this teardrop-shaped bifacial stone tool was first used by *Homo ergaster* and *Homo erectus* as long ago as 1.78 million years BP. It ceased to be used or made (except by anthropologists) by 400,000 BP, the time of the proto-Neanderthal and well before *Homo sapiens sapiens*. This technology traveled out of Africa to Asia, the Middle East, and Europe. Its uses are unknown, but it is presumed to be a multitasking tool for hacking (tree limbs); digging; butchering, including cutting and breaking bones, and scraping hides. And one anthropologist thinks that the axe was a possible projectile thrown like a discus at flocked prey. Recently some have speculated that very large versions of the hand axe—too big for practical use—may have been display objects. I know of no other technology with this long a use-span.
>
> Today's cell phone beats or ties the axe's ubiquity. Social scientists estimate close to 95 percent of today's human population has access to cell phones since undeveloped countries have them as "leapfrog

technologies" and Grameen Bank developed uses for "phone ladies," who for small fees rent them to villagers in many countries. Calling, texting, photographing, calculating, and locating are some of the multifunctions of this contemporary—only a few decades old—technology. My favorite cell phone photo is of a Masai tribesman, spear in one hand, cell phone in the other on the African veldt.

Closer to my own practices are writing technologies, and whereas clay tablet-stylus-cuneiform technologies may have lasted a millennium or more, they are now gone. Pen and brush for ink and calligraphy remain in use after six or seven millennia. But the typewriter—along with such industrial technologies as the steam engine—lasted less than two centuries.

And, as you might guess, if we look at postmodern technologies, the use-spans are shorter and shorter. The turnover for recording technologies—Edison tubes, vinyl records, wire and tape recordings, CDs to MP3s and such, all happened within a century.

Philosophies

The philosophy I shall deal with here is that which deals with science and technology, often called today technoscience. In contemporary thought this spectrum of philosophy has been highly influenced by science studies in addition to philosophies of science and technology.

The style of science-technology analysis that emerged at Stony Brook in my technoscience research seminar, to be followed here and more specifically described in the last chapter, explicitly takes into account technologies, instrumentation, and is called *postphenomenology*. It also incorporates elements of American pragmatism. John Dewey held that philosophies should deal with problems facing humans, should be instrumental and experimental, and he eschewed foundations, essences, and universals. *Question: Should philosophies, like technologies, have use-lifes?* While my own answer is a qualified "yes," I am well aware that this is not the usual way in which philosophy is thought of. For many there is a kind of ahistorical history bound to philosophy: After all, we introduce much of our philosophy to students by having them read the philosophical classics: pre-Socratics, Plato, Aristotle, the medievals, and particularly the early moderns—Descartes, Locke, et al.—and then on to Kant, and then the nineteenth and twentieth centuries. And in all this the literary conceit is that all are our philosophical contemporaries. Such ahistorical history would be very unlike science, which has a history of *disappearing scientific objects*: Democritus's hard, indivisible atoms, Aristotle's crystalline spheres, phlogiston, aether, the four humours, and most recently event horizons—all are gone except as interesting but quaint historical objects.

Science-Technology Studies

I am going to indirectly suggest that my question about philosophical obsolescence, abandonment, and forgetfulness is stimulated by the paradigm shifts concerning science-technology studies in the mid-twentieth century. I will trace a brief history of STS here:

> If we take early modern science to have originated in the seventeenth century, then by the end of the nineteenth century there was a firmly established and triumphal philosophy of science. Its main outlines included a faith in progressive accumulation of knowledge, the approximation or attainment of

universal natural laws, objectivity, and freedom from religion, culture, and values. Scientific knowledge was either the only valid, or at the least the best, form of human knowledge.

The early twentieth century was more specifically marked by interpreters—themselves usually scientifically and/or mathematically trained—who saw *mathematization* and *theory production* as the core of science. Pierre Duhem, Ernst Mach, and Henri Poincaré were the best-known names. I have argued elsewhere that Husserl, himself mathematically trained, largely bought into this interpretation of science (see Chapter 2).

Traditional philosophy, including its epistemological and metaphysical emphases, was being pushed back, with logical empiricism, logical positivism, and analytic styles of philosophy ceding to science its autonomy and rational superiority. Science produced knowledge; philosophy checked its logic and propositional form.

By mid-twentieth century a "pushback" from within philosophy of science begins. The 1950s and '60s are the beginnings of the positivist-antipositivist controversies. Thomas Kuhn, Paul Feyerabend, Karl Popper, and Imre Lakatos stand out in Anglophone circles. All called into question the elements of the progressivist, triumphal interpretation of science noted above and argued for more discontinuity, more social-technical embeddedness, and one or another "framework" interpretations.

From within philosophy of science, however, there was very little note of technologies, materiality, or social-historical factors. And whereas in Europe there were beginnings of philosophical attention to technologies, particularly in response to the industrialization and militarization of previous decades, there was not yet what could be called philosophy of technology.

Although it is clear that what I am calling the pushback begins to take shape in the mid-twentieth century, my too brief caricature above overlooked certain parallel movements in science interpretation from the social sciences. There had been a "sociology of science" regarding science that was dominated by Robert Merton and colleagues from 1957 through the 1960s. Then from phenomenology and Alfred Schutz, *The Social Construction of Reality* by Peter Berger and Thomas Luckmann was published in 1966. These alternative STS interpretations, however, begin at the same time as antipositivism and Kuhn's *Structure of Scientific Revolutions,* published in 1962.

We now have two sets of interpreters on the scene: on the one side, philosophers of science and scientists themselves; on the other, social scientists. For the first set, the philosophers of science and scientists, the preference is for theory, propositions, equations, and logic—in short, a highly "conceptual" approach to science. The social scientists emphasized what I shall call, broadly, science practice. And this emphasis creates different perspectives. First, drawing from the names already cited, I look at some of the different dimensions emerging from forefronting practices:

Robert Merton coined many of the terms that describe social behaviors and expectations within science's sociality, such as "role model" and "reference group," and noted many nonlogical features such as "unintended consequences" and "self-fulfilling prophecies."

But much of his focus became that of looking at how scientists forge conclusions and reach consensus through citation practices.

Thomas Kuhn also attended to the social practices of scientists, taking account of how, when a revolution occurred in science, beyond the arguments that took place, there tended to be a slow generational shift of groups favoring earlier compared to later paradigms.

Popper and Lakatos looked at science practice as developing research programs involving different groups. Pure logical arguments, even falsification, had limits; thus the image of science here varied from that of the positivist dominant emphasis.

And glimmers of the role of technologies in science practice begin to appear. Alfred North Whitehead in *Science in the Modern World* (1963) takes particular note of twentieth-century instrument improvement. "In science the most important thing that has happened in the last forty years is the advance in instrumental design."[1] And Kuhn notes repeatedly in his *Structure of Scientific Revolutions* how new observations via improved instruments make paradigm shifts possible.

This mid-century (1950s–60s) antipositivism was soon followed by many more radical shifts in STS studies beginning in the '70s. The first of these new social science movements were from the UK, the so-called "Bath School" with its "Strong Programme in the Sociology of Scientific Knowledge" led by Barry Barnes and David Bloor. Unlike Mertonian sociology, which looked at social structuring, the SSK program looked at how scientific conceptions were themselves socially constructed. *Scientific Knowledge and Sociological Theory* was published in

1974, followed by *Knowledge and Social Imagery* in 1976. Both openly defended relativism philosophically. In this same time period, Bruno Latour and Steve Woolgar began to "follow the scientists around," but in the laboratory, not in the theoretician's office. *Laboratory Life* appeared in 1979. Interestingly, the subtitle of the first edition was *The Social Construction of Scientific Facts,* and the second edition has only *The Construction of Scientific Facts.* Latour, the primary author, deliberately takes an anthropological-sociological participant-observer point of view in a biochemical laboratory to take account of its practices. Laboratories, he contended, are where scientific knowledge is produced. "Social constructionism" became the controversial title for a wider group of science studies analysts, including Trevor Pinch, Harry Collins, Andrew Pickering, Karin Knorr-Cetina, with publications widely appearing in the 1980s. This was, simultaneously, the same time period for the emergence of an Anglophone philosophy of technology. My *Technics and Praxis: A Philosophy of Technology* (1979) was followed by books by Albert Borgmann, Langdon Winner, Andrew Feenberg, and Hubert Dreyfus in the '80s.

What I am depicting here is the rise of multiple reconsiderations of science, coupled to an increased interest in technologies—all of which shift the understanding of both science and technology toward more historical, cultural, and material dimensions. To complete the depiction, however, I need to mention the rise of feminist critics of science and technology, which also begins to take place largely in the 1980s. The biologists Evelyn

Fox-Keller and Donna Haraway and the philosopher of science Sandra Harding stand out here—each showing how gender roles play significant parts in science practice. I cannot take the time here to review the intense battles that took place concerning STS in the late twentieth century, but the results are tellingly clear:

All parties agree that the understanding of science as acultural, ahistorical, unified, and triumphal—the view dominant at the beginning of the twentieth century—is dead. In a wide-ranging genre review of philosophy of science in *Choice* in the December 1989 issue, Steve Fuller pointed out, "There has been only [one] widely cited book to extend the positivist program since 1977."[2] Along with that, all parties now acknowledge in different ways the fallibilism of science and the demise of positivism. Larry Laudan's *Science and Relativism* (1990), in which he writes "we are all fallibilists now,"[3] and Ernan McMullin's *The Social Dimensions of Science* (1992), in which a spectrum of contributors acknowledges and examines these social dimensions, are major establishment acknowledgments of this shift.

Then, within philosophy of science, one begins to see the rise of careful analyses of, in particular, the role of technologies and instruments, including Ian Hacking, *Representing and Intervening* (1983), Peter Galison, *How Experiments End* (1987) and *Image and Logic* (1997), and Robert Ackermann, *Data, Instruments, and Theory* (1985). My own *Instrumental Realism: The Interface between Philosophy of Science and Philosophy of Technology* (1991) gives a full account of this emergence of interest in instruments.

I want to close this brief history of change in philosophy of science and STS studies with a glance at a clincher book by Ronald Giere, *Scientific Perspectivism* (2006). Giere, long an establishment philosopher of science, head of the Minnesota Center for Philosophy of Science, long associated with a positivism program, here looks explicitly at the impact of movements I have been citing. He ends up accepting the embodied and situated epistemology of these movements and develops his own visualist perspectivalism as the basis for science's observations. As Bas van Fraassen's blurb puts it: "Giere . . . has placed perspective center stage, contributing to innovations that are changing philosophy of science today as radically as the historical turn of a half century ago."[4]

Husserl's Missing Technologies—Analysis

So where does Husserl fit into this narrative? I propose here to reread Husserl retrospectively from precisely the changes in philosophy of science and STS studies outlined above. But I shall also look back at Husserl's own time and the interpretations of science then dominant. Clearly, chronologically, Husserl produced his work only into the early part of the twentieth century. To open the topic for Husserl, I begin with three generalizations about his philosophy of science and dealings with technologies:

First, as I show in the next chapter, Husserl's explicit philosophy of science remained largely within the early twentieth-century mathematizing umbra. Although the late *Crisis* begins to anticipate the turn to practice, it remains ambiguous. His treatment of Galileo remains strictly within the limits of mathematics, and although he worries about how this abstract and theory-biased approach separates science from the plenary lifeworld, he leaves Galileo as a mathematizer.

Husserl's big worry about science was that it was losing its drive for a deep rationality; he posed the development of transcendental phenomenology as a rigorous science on its own terms as, in effect, a differently aimed science.

Although it is widely known that Husserl paid little attention to technologies, the hints in "The Origin of Geometry" and the *Crisis* about writing technologies gave glimpses into his praxis-lifeworld relations. The "Origin" develops the notion that geometries arose out of concrete practices—such as remeasuring and establishing field boundaries after the annual Egyptian floods—and he uses carpentry examples from planning, which made for ever "purer" shapes toward ideal limits. I will discuss this in more detail in the next chapter. Here I focus much more narrowly on Husserl's comments on *science technologies,* or *instruments,* in particular the telescope and microscope as noted in his *Nachlass* in Husserliana 39 (2007). As mentioned, Rochus Sowa, the editor, was at the 2007 Husserl Circle and early sent me the German texts. Husserl calls this the nearby-far-off-world. And to bring this back to his view of science as well, I shall parallel Husserl on the telescope and microscope with Galileo, who, contrary to the impression given in the *Crisis*, actually had a great deal to say about the telescope, with a major set of arguments to extol its superiority to the eye. My textual sources include Rochas Sowa's German texts from Husserl's *Nachlass* (1916–37) through an English translation by Francis Bottenberg (2010) and Harold I. Brown's "Galileo on the Telescope and the Eye," *Journal of the History of Ideas* (1985).

First Husserl and what he does not do: Nowhere in the comments is there any *phenomenology of instrument use,* nor does one gain the impression that Husserl himself had direct experience of telescope or microscope use. But that would be strange, because he studied astronomy, along with mathematics, logic, and philosophy, in Leipzig, 1876–78. Indeed, to the contrary, while he calls what is seen through these instruments the *Nah-Fern-Welt,* the near-far-world, he continues to call the instruments telescope-things and microscope-things. Similarly, he calls what is seen through these instruments a *correlate* of our pure perceptual self-persisting. In passing I simply note how this contrasts with both Heidegger's sense of instrumental "withdrawal" in use, Heidegger's hammer, and Merleau-Ponty's extended embodiment of canes, cars, and hat feathers. To make the point even more starkly, Husserl's approach to such "instrument things" is the reverse of Heidegger's. It is as if one must first conceptually recognize the hammer before embodying it for praxis: "The form of a tool, its entire composition and immediately recognizable materials: the style of the hammer, its iron, heavy head; the bicycle, the form of the bicycle, the rims, the iron wheels, etc."[5] What I am suggesting is that for Husserl, tools, technologies do not lose their *objectness.* It is as if one must first look at them as things, and then adapt them to praxis. This is part of his recalcitrant modernism and vestigial Cartesianism. It plays a double role: On the one hand, the "thingness" keeps Husserl from seeing how instruments in practice differ from instruments as objects, and, on the other, it retains—as if no reductions were made—a type of prephenomenological "natural attitude" externality. In *Ideas II,* an early text, this vestigial Cartesianism occurs in a rare discussion of coal! The section in which this reference occurs is on the person and his surrounding world. While the deep phenomenology of the section clearly indicates the interrelatedness of personal *Ego* and surrounding world, Husserl situates the first take on world as an awareness of objects (coal as one), intimating that coal is first an object with "objective" properties. "To be hot is an Objective property. . . . This object is 'combustible' material (at first without any practical bearing)."[6] Indeed, in the textual context—which is ambiguous, as I shall show—it is as if objects are first given as objects, only later to become valued, useful, beautiful, and so on. "Afterwards, the objects become apprehendable as in the service of the satisfaction of such needs according to this or that property."[7]

However, in a broader context it is unclear whether Husserl is writing in what is obviously the common Cartesianism of the time or whether he is on his way to an inversion of a natural attitude (which is referred to regarding the physics of causality later in the chapter). Husserl does allow that one may immediately apperceive a violin as "beautiful," but for the most part all values, uses, pleasures, and so on are "added" on. This is a tactic that is diametrically opposite that of Heidegger, who derives "objectness" from broken, missing, or absent usage.

Now I switch to Galileo: Husserl's Galileo has no telescope, but part of the blame rests with Galileo himself. Galileo in his most "philosophical" texts remains the theoretician-mathematizer. Galileo repeats the distrust of the senses of his times; he enhances the mathematical language of the "book of nature" and, as we shall see, with his strong preference of telescopic vision over the eye, continues this prejudice of early modernity. But in terms of the history of science, all four of Galileo's primary astronomical discoveries were due to his telescope. And he immediately bragged about his sightings, claiming that neither Aristotle nor the Bible saw such sights. These sights include the mountains and seas of the moon—which admittedly he also measured with crude shadow calculations to find that these were as high as the Alps; the phases of Venus, and the satellites of Jupiter, together finalizing his shift to Copernican heliocentrism; and sunspots, which got him into his first troubles with the church. (No European then knew that the Chinese had, more than six centuries earlier, already observed and clocked their cycles.) All Galileo's astronomical discoveries were made possible only via his telescope. And although Galileo did nothing like a phenomenological analysis, he wildly praised telescopic vision and conducted a whole series of visual comparison experiments between telescopic and eyeball vision.

Galileo began to telescopically observe the heavens in 1609–10, and he quickly began to extol his discoveries in *The Starry Messenger* (1610). While there was a cultural consensus regarding suspicions of the senses, there were also doubts about the then usually amateur lenses (Galileo made his own) which could also produce distortions, instrumental artifacts, and double or triple images. Galileo was himself well aware of these limitations and cautioned the Medici court that perhaps they should not attempt to observe Jupiter's moons unless he could be present himself.[8]

The background for all these observations was, of course, the contest between Ptolemaic geocentrism and Copernican heliocentrism, since degrees of brightness, disc sizes and phases would be different in the two solar systems. Galileo describes these problems in his *Dialogue on the Two Chief World Systems* (1610):

> For if it were true that the distances of Mars from the earth varied as much from minimum to maximum as twice the distance from the earth to the sun, then

when it is closest to us its disc would have to look sixty times as large as when it is most distant. . . . No such difference is to be seen. Rather . . . it shows itself as only four or five times as large as when, at conjunction it becomes hidden behind the rays of the sun.[9] Similarly Venus, when it is beneath the sun and very close to us, its disc ought to appear to us a little less than forty times as large as when it is beyond the sun and near conjunction.[10]

These are predicted eyeball observations that do not happen, but surprisingly, Galileo argues that the telescope corrects what is a *defect of vision*: "These things can be comprehended only through the sense of sight, which nature has not granted so perfect to men that they can succeed in discerning such distinctions. Rather, the very instrument of seeing introduces a hindrance of its own."[11] Galileo contends that naked-eye vision comes with "adventitious rays" or an aura that makes planets look larger than they are. The telescope not only magnifies but also eliminates this distortion. But even with the telescope one celestial anomaly remained. Although planets presumably appeared as correctly magnified without "eyeball rays," "stars, whether fixed or wandering, appear not to be enlarged by the telescope in the same proportion as that in which it magnifies other objects."[12] Here we have a major clue to the limits of early modern astronomy. Galileo, no more than Ptolemy or Copernicus, had any idea about the vast size of the universe! The stars were simply so far away that their disc sizes did not change with an optical telescope!

I shall go not much farther with Galileo. Brown notes that Galileo did undertake a lot of experiments to show that the adventitious rays were in the eye and not from the light source. These including pressing on the eyeball, using ropes and black and white strips of cloth to block off images and others. As noted, Galileo lived in a time in which senses were distrusted, but in his arguments for the superiority of telescopic vision two further moves are implied: First, Galileo was treating the eye itself as an *instrument*. He was not yet under the later notion of the body as machine, but his willingness to be critical of human perception as having structural defects was part of his early science. He was aware of flaws in the lens grinding of the time and of the various instrumental artifacts these produced, but he was unaware of many of the technical limitations that would face optical development. The "chromatic distortion" is one of these— because lenses are curvilinear and the colors within "white light" have different frequencies, it turns out that at approximately 30 power this distortion takes place unless the lens is of flint glass. Galileo's most powerful telescopes were close to 30 power, not quite enough to resolve the proper image of Saturn's "protuberances" as rings but not yet to the chromatic distortion limit.

Now I return again to Husserl: As I noted above, Husserl places things (with predicates) as what is immediate: "The completely near, the true. [The nearby-far-off-world, that which is instrumentally mediated by the microscope and telescope, is a] correlate of pure perceptual self-persisting."[13] He describes this correlation as a *relativity*. This relativity, this correlation, however, is part of the *unity* of experiences. Thus the introduction of instruments is the introduction of a *new stage*. My suspicion is that this is parallel to the stages in the "Origin of Geometry" where the concrete measuring practices stimulate or lead to the more abstract stage of geometry. "Every true determination in the earlier stage of experience is preserved in the later one. . . . This is a sequence of stages towards the exhaustive experience of the same."[14] And, as with the *Crisis*, what is primary is a here-and-now bodily experience: "The experiencer has at every moment of experience a sensuous visible world, oriented around his body [*Leib*]. He has a spatial and temporal present around his now and his here."[15] At this point in the *Nachlass* Husserl goes into something of a detour into how historical knowledge is included into unity with the here and now present—through empathy—and then *sedimented knowledge* by which he connects instrumental and eyeball vision. "Perceiving the moon, someone might say: I see the lunar mountains."[16] But although we do not perceive lunar mountains except for the learned mediation of the telescope, this is a report of sedimented knowledge. One can compare both sides of the correlation: "The telescope-thing references the thing of the naked eye and up close. . . . [Thus through this comparison] spots on the moon, I learn, are mountains."[17] We are, as with the coal example, adding on knowledge.

Here, I interrupt: Husserl's description of comparing eye and telescopic sightings of the moon do not seem to me to be either phenomenological—or anything like my own experience of telescopic sight—or of Galileo's experience! Indeed, reading Husserl makes me think he had not himself experienced a telescope or microscope, although in his Leipzig studies, he must have done so. I do have a reflector telescope in my Vermont place, and we occasionally hold "telescope parties" with guests coming to see precisely what Galileo saw in 1609 and 1610. The moon comes first—and without exception, we, our guests, harking back to Galileo, have the "aha" experience. There is an immediate recognition of the moon's mountains and craters. And simply by varying from telescopic to eyeball sights, whatever sedimentation there might be, the sameness of the moon seen differently is immediate. Moreover, the experienced shift in space-time between eyeball and telescope is also immediate. And the relative magnification of moon and my bodily motion through the telescope is also immediate. This is to say that the experience *through* the telescope is not primarily *of* the telescope. The telescope becomes experientially dominantly transparent, and the target or reference of the telescopic experience is the moon! The telescope becomes the "means" of a new moon experience. The immediacy of moon mountains through the telescope is a mediated *perception*. And as any phenomenology of perception shows, it is interrelational.

Husserl does not describe this "aha" transparency, nor does he notice the "withdrawal" of the telescope in use. But here is something worse. On the original occasion of the presentation of this chapter at the Husserl Circle meeting in Dartmouth, June 2014, there was a photo exhibit on Husserl, and, as noted, almost all photos showed him with his distinctive eyeglasses. Were these "eyeglass-things"? And, if so, is what he sees with his eyeglasses not his colleague, Landgrebe, but a perceptual correlate? I am unaware of any Husserl phenomenology of eyeglass vision, but as an optical technology it poses the same embodiment phenomenon as more complex optics such as telescopes and microscopes. To be sure, the near-far is less pronounced, but eyeglass users, when getting a new prescription, undergo a time of having to adjust to a different set of bodily motilities than from before, or from experience without glasses. One quickly relearns how to be normally motile, and this quickly learned embodiment becomes familiar and sedimented. This problem again reflects the vestigial Cartesianism to be found in Husserl. And for science, this lack of a phenomenologized technology creates the problem of distancing the lifeworld from a world of science. This is to say that at least whenever Husserl began to need eyeglasses, his lifeworld became one mediated through an optical technology.

Returning to Husserl on the telescope and the microscope, for Husserl these technologies are machinic. Yet, although machinic, Husserl does end up giving a positive evaluation to these technological means: "It is here that the invention of machinic means for heightening the power, the breadth, and generally the exhaustiveness of natural experience (microscope, etc.) plays a role, as it amounts to an improvement, an enhancing configuration of naturally growing organs." Here we move closer to Galileo. Finally, then, Husserl gives a virtual accolade to technologies as instrument. I will include a rather long quotation:

> Human beings can be divided into those who have stayed behind (deficient experiencers) and those who have advanced; and the advanced now represent the norm, they are subjects of a relatively more determinate, although unchanged world. Thus the world as seen through the telescope and the microscope offers more truth than the world that previously counted as certain in its being and its lawfulness.[18]

So, with Husserl, we now have stages, and stages that can improve our knowledge of the world. This means that compared to Galileo, Husserl does see instrumentally improved knowledge as an improvement, but unlike Galileo he is not downgrading perception nor calling it illusory. Yet as he reaches his conclusions, he does something even more radical in a move that anticipates that "we are all fallibilists."

> When the development of experience, or of humanity, uncovers the prospect of new stages of experience continually being possible and attainable through future inventions, *there is no longer a world that is given or could be given as ultimately true through unified perception. . . . The truth of the "sensuous" world now necessarily contains a relativity towards "sensuousness," towards the stand of exhaustive experience. We have an open infinity of universal world truths before us . . . No individual determi-*

*nation can be final [yet] to heighten the level of
experience, to improve human organization itself
through technical means and thereby to uncover a new,
wider world but the same as the earlier one, but contains
richer and better truth, and thereby to make a richer
world praxis possible.* (Emphasis mine).[19]

This is a surprising conclusion, and it turns out to be, I contend, prescient with respect to what happens within technoscience in a postmodern guise. As the history shows, both
Galileo and Husserl remain within what could be called the
"first revolution" in sciences with technologies. But both
draw different comparisons: Galileo clearly strongly differentiated between the bodily "instrument" of the eye and
what he took to be the improved instrument of the telescope.
Husserl, in contrast, clearly took the up-close here and now
as "truth," which could be extended through the "correlation"
of telescope-things. (Thus Husserl implies a position concerning technology that would have it as a kind of "applied"
science.) But, in relation to the history of science and instrumentation, both for the most part remained within the
limits of optical astronomy. This astronomy remained limited to what might be called "analogue" or isomorphic perceptual constraints. The eye-telescope comparisons remain
easy and limited to our perceptual horizons despite the
differences between what is seen and how it is embodied
through instrumental-mediated experience.

Husserl did live into the very beginnings of what I am
calling postmodern science, in which new instruments began to reveal dimensions of reality *no longer* isomorphic or
analogue to our bodily perceptual experience. But he lived
just barely into this time. In astronomy Hubble did use a
massive 100-inch telescope to proclaim Andromeda a second galaxy in 1924—heretofore only the Milky Way was

recognized as a galaxy. Then, for example, the first radio—nonoptical—telescope was a crude radio-antenna invented by Karl Jansky of Bell Labs in 1931. It could detect directionally radio waves from storms (heard as static) and even the hiss that was to be identified much later as the background radiation of the universe. But the first more precise paraboloid radio-telescope was not invented until 1937 by Grote Reber, one year before Husserl's death.

Then if we jump from this early twentieth-century extension of astronomy beyond optical imaging to the present, today's imaging brings instruments that can map the entire electromagnetic spectrum and have led to a universe *radically different from both Galileo's and Husserl's worlds.* Neutron stars, rapidly rotating pulsars with polar jets, black holes, multiple galaxies of many shapes, even gas clouds and solar flares did not "exist" for science prior to the mid-twentieth century. Here is Husserl's "open infinite of universal world truths" brought into being through new instrumental technologies.

So, I close with a question posed earlier: Should not philosophies, as with sciences, also be prepared to change, shift gears, and find ways to deal with a new, wider world? I am suggesting that classical phenomenology, with Husserl, remained within the analogue-isomorphic limits of instrumental technologies and that new strategies for dealing with the worlds revealed beyond our bodily limits are now necessary.

2

Husserl's Galileo Needed a Telescope!

This chapter is a look at Husserl's explicit philosophy of science in the light of contemporary analyses of science in practice. The *Crisis,* published in 1936, was his last major publication on this topic. Yet it takes very little imagination to realize that since 1936, epochal changes have occurred in both the sciences and the interpretations of science, including philosophy of science.

Twentieth-century history of science, I argue, has been marked by changes that are at least as profound as those that marked the turn from Aristotelian science to early modern science in the sixteenth and seventeenth centuries. Since World War II, the sciences have increasingly become Big Sciences, the term for which was coined by Derek de Solla Price in a book of that name (1962). Late modern science— first with chemistry, then with physics, engineering, and today the biological sciences—has become a science of corporate groups, large state funding, complex technologies in instrumentation, and implied major social-political dimensions for its operations. Husserl did live long enough to see and to some extent appreciate that in his recognition science had become "research."

Conceptually, there is as much distance between Newton and Einstein–Bohr as earlier between Ptolemy and Copernicus. The interpreters of science—at first primarily historians and philosophers—began to catch up to these changes in the 1950s and 1960s, well after Husserl's death. The move away from science as theory machine, and the earlier notions of unified science, verification, and universal law occurred with the "new philosophies of science" proposed by Thomas Kuhn, Paul Feyerabend, Karl Popper, and Imre Lakatos. And they too to some extent were propagating a notion of science as a research process that engaged corporate structures, vast technologies of instrumentation, megafunding, and the major social-political role that characterizes Big Science, as well as recognizing science as no longer the unified, generalized process early twentieth-century philosophers took it to be.

Nor did the "new" philosophies of science have the last word. The emergence of a virile set of new social sciences perspectives on science, perspectives that focused on science praxis, laboratory life, and the instruments or technologies of science also began to flourish later in the twentieth century.

Pre–World War II philosophy of science was simply superseded by the late modern turns of post–World War II science and its now multiple interpretations. Thus, we can expect that Husserl's philosophy of science will turn out to be, in significant respects, part of "history."

Where Was Husserl?

By the question "Where was Husserl?" I mean to locate both in historical time and from Husserl's chosen perspectives the outlines of how Husserl interpreted science and philosophy of science. Biographers and historians interested in Husserl know that from his own background and discipline, logic and mathematics were the primary cognate disciplines mastered by Husserl. And so far as the sciences go, geometry and physics, with very small asides to astronomy, gain the most concentrated attention. The same may be said of his social engagements with scientists and philosophers of science of his day. The Göttingen group is an obvious standout, as were the mathematicians around David Hilbert.

What kind of sciences are these? First of all, they are the most abstract of sciences, the sciences most given to formalisms, mathematization, and idealization. Second, they are the least technologized of the sciences (excepting physics, of which more later) at least in the early twentieth century. Third, they are the sciences that least directly engage the whole body, or embodiment practices.

And fourth and finally, they are the sciences that appear to be the most ahistorical insofar as they are expressed in the special language of mathematics, which must be learned in special ways by all cultures approaching mathematics.

Yet these were the paradigmatic sciences in the background for Husserl and for his symbolic spokesman for early modern science, Galileo Galilei (one could add that not only is this what Husserl sees when he talks about the sciences, but that this is also what he himself knows best in terms of science praxis).

Before turning to what is revolutionary and radical in Husserl's philosophy of science, I shall first take a few perspective glimpses of the state of the art during his productive apogee. By first turning to an abbreviated history of twentieth-century philosophies of science, one can better locate Husserl's site.

Twentieth-Century Philosophy of Science in Three Steps

Early twentieth-century philosophies of science focused clearly on what philosophers saw as the mathematization process, which they found paradigmatically in the disciplines of physics and, in a more concrete sense, in astronomy, both of which were in effect a refinement of early modern epistemologies of the so-called geometrical method.

Mathematizers. At the time of Husserl's early thoughts on Nature, for example in *Ideen II,* the three most prominent philosophers of science were Jules Poincaré (1854–1912), Pierre Duhem (1861–1912), and, particularly for Germany, Ernst Mach (1838–1916). Among these, there was a nuanced consensus about the nature of knowledge and the role of mathematization. (a) First, all held to the basics of early modern epistemology, which has its subject-in-a-box (body) viewing its sensations in a way that can infer an external world only through representations. (b) This in turn leads to the notion that science is physical theory, and its scientific exemplar is physics. The reasons for this narrowing of the sciences include the fact that most of those doing philosophy of science were themselves physicists or philosopher-physicists. Additionally, given the theory-bias of early twentieth-century physics, physics as a discipline was more inclined to self-referencing. (c) Third, theory takes its shape through mathematization, whether this process is considered as a formalism (Duhem) or as instrumental (Mach and Poincaré). Thus science is the process of mathematizing the world through theory and is paradigmatically exemplified in mathematical physics. Husserl, very much part of this milieu, thus also characterized science in its essentially physics mode, in a formalistic and abstract way. This was later to become part of what in the *Crisis* was characterized as science's forgetfulness of the lifeworld.

I think it is rather easy to see that it is precisely this "standard view" that Husserl himself takes for granted in his own characterization of at least the way science takes itself (in the early twentieth century). It was a rather "metaphysical" view of science, highly generalized and virtually nonempirical. Part of my thesis here is that Husserl is simultaneously both highly conservative and yet also radical with respect to his view of science. This taking for granted of the then standard view of mathematized physics as standing for science overall (*überhaupt*) is Husserl's conservative moment.

Positivists. The second movement in the philosophy of science, was logical positivism or empiricism, centered in the Vienna Circle in the 1920s. Within this movement, the consensus included four points. (a) The notion of a unified science was that all sciences, ideally, would ultimately be unified in a single model. But such a unified science would also be reductionistic; ideally all other sciences would refer to or be subsumed under such a physics. (b) A unified science would also be that which could progress through a hypothetical-deductive process, whose logic and propositional structure would be the task of philosophers of science to clarify and outline. (c) Once again mathematized physics was to be the favored model for operational sciences. (Those of you who are historically aware may recall that in the process of "purifying" science in this way, some members of the Circle in effect decided that such fields as geology could not be a genuine science since it did not well follow the hypothetical-deductive model!) The one addition that logical positivism or empiricism did make to early twentieth-century philosophy of science that would place it necessarily closer to Husserl, was the addition of (d) a verification process whereby observations must play a much more important role in science per se. Observations implied a more important perceptual role than had been taken in the earlier mathematized "pure" science.

Husserl himself saw the parallelism with positivism in proclaiming that phenomenologists are the genuine positivists.

> If "positivism" is tantamount to an absolutely unprejudiced grounding of all sciences on the "positive" that is to say, on what can be seized upon *originaliter*, then we are the genuine positivists. In fact, we allow no authority to curtail our right to accept all kinds of intuition as equally valuable legitimating sources of cognition—not even the authority of "modern natural science."[1]

His view of science remained clearly that of a unified science, science *überhaupt*, and not that of some set of discrete and often radically distinct praxes called sciences. And at the core of science remained the trajectory—which in his own way he questioned—which was dominantly theory driven, driven by the ascendancy of abstraction, formalism, and idealization. Husserl's philosophy of science, however, was also a radical variant compared to positivism. The notions of lifeworld and of the bodily-perceptual plena were not the same as verified sense data.

Antipositivists. The third moment in my quick three steps through philosophy of science, occurs in the mid-twentieth century and begins what is usually now called the "positivist–antipositivist" controversies. The antipositivists include, preeminently, Kuhn, but also Feyerabend, Lakatos, and Popper. Oversimplifying, one can see in retrospect that each in his way (a) introduced much stronger stands of discontinuity into the interpretations of science, whether these were historical revolutions or paradigm displacements (Kuhn), or discrete research programs (Popper and Lakatos), or a series of ad hoc procedures not marked by unity (Feyerabend). (b) These antipositivists also began to introduce more history as such into the image of science, and to a lesser extent, particularly with Kuhn, some sense of the role of instrumentation. (c) But one could continue to claim that the antipositivists retained the dominance of theory-production as the core of science activity.

Since Husserl was no longer alive by midcentury, any relation of his work to this third movement has to be by way of presumed anticipations.

Husserl in All This

If one can speak of Husserlian hermeneutics, one would always find it circuitous. Husserl begins with some set of taken-for-granted beliefs, in this case the mathematization view of science, in which the turn to a fully quantifiable analysis of nature constitutes the essence of science itself. He then circles backward and downward and asks what it is that science itself takes for granted or must hold implicitly. In the *Crisis* this is the unique and peculiar form of historical derivation from a lifeworld to any particular science. Husserl's process is a "deconstruction" followed by a "reconstruction." But in order both to foreshorten this process and to invert it, let us begin with the conclusions that Husserl claims to have reached by his questioning back to origins.

First, what is basic, necessarily presupposed, and titled as "prescientific" is a lifeworld that is (a) bodily-perceptual:

> In the intuitively given surrounding world . . . we experience "bodies"—not geometrical-ideal bodies, but precisely those bodies what we actually experience, with the content which is the actual content of experience.[2]

(b) It is filled with practices. I am taking as praxis the activities that constitute various meaning-regions, such as Husserl's "invention" of Egyptian surveying praxis:

> The art of measuring discovers practically the possibility of picking out as [standard]measures certain empirical basic shapes, concretely fixed on empirically rigid bodies which are in fact generally available and by means of these relations which obtain . . . between these and other body-shapes it determines the latter intersubjectively and in practice univocally—at first in narrow spheres (as in the art of measuring land). . . . The art of measuring thus becomes the trail blazer for the ultimately universal geometry and its "world" of pure limit-shapes.[3]

(c) The lifeworld is cultural—historical:

> The geometry which is ready-made, so to speak, from which the regressive inquiry begins, is a tradition. Our human existence moves within innumerable traditions. The whole cultural world, in all its forms, exists through tradition. . . . Everything traditional has arisen out of human activity, that accordingly past men and human civilizations existed, and among them their first inventors, who shaped the new of our materials at hand, whether raw or already spiritually shaped. . . .
>
> For a genuine history of the particular sciences, is nothing other than the tracing of the historical meaning structures given in the present, or their self-evidences which underlie them.[4]

All questioning back simply arrives at this foundation. Thus the most general claim which can be made about science is that it arises from, is dependent on, and presupposes a perceptual, praxical, historical lifeworld

Here is again something confusing: every practical world, every science, presupposes the life-world: as purposeful structures they are contrasted with the life-world, which was always and continues to be "of its own accord." Yet, on the other hand, everything developing and developed by mankind . . . is itself a piece of the life-world.[5]

Put into contemporary language this is equivalent to saying that science is not ahistorical, noncontextual, but rather is thoroughly historical, contextual, and cultural. As we shall see later, this places Husserl squarely in the midst of the current discussion swirling within the "science wars."

In addition, how is science different from the lifeworld? If it is different, does that imply a distance from the lifeworld? Here is a nexus of so much of the problem that arises for phenomenology in relation to the philosophy of science. The standard misinterpretation must be laid right at Husserl's door. It comes from his version of the symbolic first modern physicist, Galileo. Galileo, in inventing modern science forgets the lifeworld and its perceptual, praxical, historical origins and distances and substitutes an idealized, abstracted world of mathematics for the foundation. The distancing of science from the lifeworld is a "forgetting" of its foundations and a "substituting" for it a "scientific world." The complexity of the lifeworld contrasts with the ideality and abstractness of the scientific world. It is here that science is described as an increasing process of abstraction, formalization, and idealization, which leaves behind the plenary, perceptual, material bodies of the lifeworld.

Husserl's counter to this movement is his "invention" of a praxical world of geometry. In the origins of geometry, scientific idealization is the process of forgetting that the abstractions are abstractions from concrete practices of measurement—with instruments and so on. That is what presumably lies in the perceptual, praxical, historical "pre-scientific" lifeworld of geometry.

> But we now must note something of the highest importance that occurred even as early as Galileo: The surreptitious substitution of the mathematically sub-structured world of idealities for the only real world, the one that is actually given through perception, that is ever experienced and experienceable—our everyday life-world.[6]

There is an upward, slippery incline of approximations into an ideal world, which distances the investigator from the bodily-materiality of the lifeworld.

To be fair to Husserl, however, one has to note two qualifications that maintain a possible reconnection to the life-world. The first is that the very process of idealization and distancing from a primal lifeworld is not simply negative but is the attaining of a new level and region of autonomy, which is positive so long as its origins are not forgotten.

> [The confusion of the life-world and science as a piece of the life-world] is only confusing because the scientists, like all who live communalized under a vocational end . . . . have eyes for nothing but their ends and horizons of work. No matter how much the life-world is the world in which they live, to which even all their "theoretical works" belong, and no matter how much they make use of elements of the life-world, which is precisely the "foundation" of theoretical treatment as that which is treated, the life-world is just not their subject matter . . . and thus is not, in the full survey, the universe of what is.[7]

And second, because the lifeworld cannot simply be replaced, the process of idealization is characterized by Husserl as an indirect mode of analysis. It is an "application" rather than the uncovering of a platonic intrinsicality within the object analyzed.

In terms of the history of the philosophy of science, this Husserlian hermeneutic has been both the source of much pain for subsequent phenomenologists in this field and yet also the location of Husserl's radical, as compared to his conservative, dimension. One common misunderstanding that has doomed post-Husserlian phenomenology to unbased attacks concerning phenomenology's subjectivism is that phenomenology deals only with a prescientific, sensory, or perceptual world, whereas science deals with the scientific or objectively constituted world. Rather, Husserl, as Merleau-Ponty after him, was after the descriptive analysis of the very distance and constitution of the lifeworld in relation to the world of science and thus must deal with both the pre-and postscientific worlds. Yet having proclaimed the prescientific lifeworld as fundamental, it also set this mode of analysis on a trajectory radically different from the early standard views of science. Husserlian philosophy of science, well before the science wars, was dubbed "antiscientific," in part because it does not presuppose the automatic priority of a "scientific" world.

But its radicality, too, depends on the role that praxis has in the very building up of a science.

The problem lies in the various distinctions and relations between the lifeworld and the worlds of science. From Husserl's point of view these include three things. (a) The relation of foundation to derivative or result. There can be a lifeworld without science, but there can be no science without a lifeworld.

The life-world is the world that is constantly pregiven, valid constantly and in advance as existing, but not valid because of some purpose of investigation, according to some universal end . . . scientific truth presupposes it . . . and in the course of [scientific] work it presupposes it ever anew, as a world existing in its own way.[8]

(b) Then there is the relation of the general "whole" of the lifeworld—it is presupposed by any human praxis whatever—and the partial or specially autonomous practices of science. Its abstractness implies that it remains partial. (c) But the special and autonomous situation of such practices are nevertheless both differentiated from the lifeworld and in a particular sense "transcendent" to it. And it is in relation to this last relation and distinction that Husserl places the role of mathematization.

Husserl Got It Wrong

This characterization of Husserl's understanding of lifeworld/science relations and distinctions in some ways sounds distinctly contemporary. To this point I have characterized only three moments in twentieth-century philosophy of science. But by midcentury interpretations of science began to emanate from what was to become known as science studies. Here the canon includes figures such as Steve Shapin and Simon Schaffer, David Bloor and Trevor Pinch, and Donna Haraway and Bruno Latour. These contemporary historicizers, sociologizers, and anthropologizers of science, that is, the science studies folk, would certainly agree with Husserl that science presupposes and remains cultural, historical, and anthropological all the way down. They would also sharpen Husserl's insights about the particularity of science practices by no longer leaving these at the level of a "metaphysical" abstraction as Science with a capital letter. There are, instead, simply a series of different sciences in the plural, related at best by "family resemblances." And contrary to the reductionist philosophers of science who portrayed science as a logical–mathematical theory machine, contemporary science studies see the sciences as multidimensional practices that include an internal culture, embodiment in machinery, and expressiveness in special "tribal languages," only part of which is mathematical. With not too much stress, a lot of Husserl could be made to fit better with this image of the sciences than he could with his own philosophy of science peers.

But—as I have indicated in my subheading—in a very serious way he also got it wrong, or he got science itself wrong and that through his own reductionistic version of Galileo. With this claim we reach the turning point and give Galileo back his instrument—the telescope.

Galileo with His Telescope

What follows will not deny the importance Galileo attached to the "mathematical language of nature" that he claimed.

> But the book cannot be understood unless one first learns to comprehend the language and to read the alphabet in which it is composed. It is written in the language of mathematics, and its characters are triangles, circles, and other geometric figures, without which it is humanly impossible to understand a single word of it; without these, one wanders about in a dark labyrinth.[9]

Nor will it deny that in his own self-interpretation, Galileo fit into the distrust of the senses that was rhetorically popular in his time. But I wish to ask a set of questions about what differences—and in particular what differences are made with respect to the lifeworld—occur if Galileo has a telescope.

First, let us turn historical in a stronger sense than the history practiced by Husserl, historical in a more historiological sense:

> We know that at late as 1597, Galileo is defending a Ptolemaic version of cosmology, a geocentric system of circular orbits.
>
> But by 1609 he has heard of Lippershey's compound lens telescope from a Jesuit friend, and given Galileo's own lens-grinding skills, he reinvents his own version of a compound lens telescope, Galileo's 9-power, which immediately improves on Lippershey's mere 3-power. (By the time Galileo quit making telescopes, some one hundred of them, he had improved to 30-power, which turns out to be the limits of lenses without chromatic distortions.)[10]
>
> And although he took quite some time before he turned the new device to the heavens, once he did, he began to flood the world with his discoveries. The four brand-new "firsts" which are usually credited to him include (1) mountains and craters on the moon, the sizes of which he estimated and which turned out to be taller than the Alps with which he was familiar; (2) the phases of Venus; (3) the satellites of Jupiter, both of which were crucial as confirmations of Copernican, which he henceforth defended, as opposed to Ptolemaic cosmology; and (4) sunspots about which he wrote and which gained him his first warning from the Inquisition.

The standard textbook history credits telescopic observations with the first observational confirmations of Copernican theory and Galileo as the holder of this honor. But, more, Galileo—never one to be humble—immediately proclaimed these new knowledges by creating the first science magazine, *Sidereus Nuncius*, usually translated as *The Heavenly Messenger*, which claimed sights not seen by Aristotle or the Church Fathers—and by extension, the Bible. This, too, would get him into trouble. But it was, of course, the mediated and magnified vision that radically differentiated his vision from that of Aristotle.

And the textbook history goes on to credit Galileo's observations with the destruction of the Aristotelian terrestrial/celestial physics distinctions by making Galilean physics "universal" or operative on both terrestrial and celestial planes.

Now, four centuries later, Galileo's telescope turns out to be both more radical than it was thought to be in his own time, but also less radical in terms of discoveries. The Jesuits, not without reason, argued that since so many telescopes provided double and even triple "images," much of what Galileo claimed was simply "built into the instrument," as what we today call an "instrumental artifact," and thus could not be trusted.

> Father Clavius . . . laughing at Galileo's pretended four satellites of Jupiter, said he, too, could show them if he were only given time "first to build them into some glasses."[11]

Galileo himself was aware of this problem, and by claiming that any man could see what he could see, and in the process see more than any ancients, also noted that only after the observer was taught by Galileo could this result be assured. And, as firsts, Eurocentrists will probably be disappointed to learn that the scandalous sunspots were actually sighted and described as early as the ninth century by the Chinese. (How? One cannot directly observe sunspots since both magnification and light-intensity blocking must occur. But the first Chinese lenses were made from dark quartz and that is how they saw the sunspots—their first eyeglasses thus were also the first "shades.")

If science is that human practice that yields new knowledge and knowledge that exceeds what can be known by the bare or unaided senses, then only the science that is embodied in instruments that amplify and magnify ordinary capacities will qualify, and that is the science produced by Galileo-with-his-telescope. And from even this scanty history, one should be able to see that much of the Galilean invention of early modern science took place only because Galileo did have a telescope, an artifact that receives no Husserlian mention.

Husserl's Ahistorical Galileo

If we now return to Husserl's Galileo and, better, to Husserl's claimed history of his forgetfulness of the lifeworld, we can see that Husserl really did not need a history at all to make the claims he made about Galileo's forgetfulness. What Husserl tries to reconstruct—presumably historically—is what Galileo had to take for granted concerning mathematics in order to invent the new science he was claiming. And what was taken for granted was the set of attainments which, originally based on praxis and perception, become idealized and taken for granted in mathematical procedures that forget their praxical basis. What is sedimented, in other words, is the second order mathematical praxis that now leaves out of consideration the perceived, plenary bodies and objects we originally encounter and that leaves out of account the attainments through measuring practices. The same points are made by Husserl in his "invented" and fanciful "histories" of geometry, which originate in Egyptian survey practices or in Husserl's carpentry examples of creating ever smoother surfaces and shapes rather than in Galileo's actual history. Husserl's fanciful "histories" are imaginative variations that distance the multidimensioned plenary perceptual objects from the idealized and abstracted secondary objects of geometry and mathematization; Husserl calls the distance one of "forgetfulness." The "history" of this process is a history of the acquisition of meaning. But Galileo's perceptions and practices with and through the telescope are precisely the measuring actions and perceptions that constitute the new "scientific" astronomical world.

But Husserl's "histories" are not histories in a second sense as well. If and when the Egyptians undertook surveying practices to establish field shapes, they did so with the praxical engagement with artifacts and technologies, just as when Galileo experimented, he engaged inclined planes, swaying pendulums, and telescopes. But these material entities remain likewise absent from Husserl's "histories." Husserl's Galileo still needed a telescope but now also an inclined plane and a swaying chandelier in the Pisa Cathedral.

Because Husserl Forgets Technologies, His Galileo "Forgets" the Lifeworld

One of the hermeneutic techniques I have developed over the years of reading our godfather's texts is that an author's illustrations not only embody the theories being developed but they also often show the prejudices involved. For example,

I challenge anyone to find a modern technology extolled, praised by, and extended in a positive way of revealing Being in the writings of Heidegger! Peasant's shoes, hammers, stone bridges, Greek temples, jugs, windmills, and hand-held pens—all serve positive and deep purposes. But steel bridges, nuclear plants, typewriters, and hydroelectric dams are all bad. In the face of this symptomatology, I find those defenders of Heidegger who claim he isn't actually against modern technology hard to believe. What kinds of examples do we find in Husserl? First, psychological, particularly perceptual-psychological examples abound to good use: Listening to musical tones, tactile examples from the hand, memory examples of protension and retension, certain visual examples—all abound to good if usually brief purpose. Then, there are the objects-before-one examples, and the already noted carpentry examples of smoothing and shaping. But—and this is a heuristic question—where are the instruments? The tools? The artifacts that are productive of change, insight, or transformation?

His Galileo fits into this symptomatology: His Galileo is not the lens grinder, the user of telescopes, the fiddler with inclined planes, the dropper of weights from the Pisa Tower, but the observer who concentrates on, on one side, the already idealized "objects" of geometry and, on the other, the plenary ordinary objects that are before the eyes but indirectly analyzed into their geometrical components. Husserl's Galileo lacks the very mediating technologies that made his new world possible.

Writing

I am contending that Husserl's Galileo is a preselected and reduced Galileo, a Galileo without a telescope, and were this

Galileo to have had a telescope there could have been a radically different analysis of science and the lifeworld. And at one point, Husserl himself came close to an insight that would have made a different analysis possible. It lies in the very brief remarks Husserl made about the "history" of writing.

The importance of written documents is that they make communications possible without immediate or mediate address from one person to another; it is, so to speak, communication become virtual. Written signs are, when considered from a purely corporeal point of view, straightforwardly, sensibly experienced; and it is always possible for them to be intersubjectively experienced in common. (Husserl here contends that writing sediments meanings, and as all sedimentation, the presentation is thus passive, but in reading, the reawakening of signification is an activation.) Accordingly, then, the writing-down affects a transformation of the original mode of being of the meaning-structure, within the geometrical sphere of self-evidence, of the geometrical structure that is put into words. It becomes sedimented, so to speak. But the reader can make it self-evident again, can reactivate the self-evidence.

This is perhaps as close as Husserl comes to identifying a material technology and its praxis playing a role in which meaning-structures are not alienated from lifeworld praxis.

It would be stretching Husserl to claim that this is much of a recognition of technological artifactuality in a mediating role in any very detailed way, but at least there is a recognition that materiality can, through its very material transformation, make meaning-structures available to bodily humans. But in Galileo's scientific praxis, this is exactly what the telescope does.

The Telescope as Mediating between Science and the Lifeworld

Let us begin with a rather close comparison between Husserl's version of writing and the Galilean telescope. First, what Husserl recognizes is that linguistic meaning-structure can be embodied and is embodied in actual speech. Sound is sensibly experienceable, and its meanings can be intersubjectively understood. But writing transforms the bodily presentation to a visible mode and to one which is "fixed" or repeatable. In this sense it is doubly "materialized." The telescope, in contrast, does not transform the sensory modalities—one can alternatively see the moon with the eyeball or with a mediating telescope. And at least before drawing or photography, which also belongs to science's trajectory, it does not "fix" its object the way writing does. But, the magnificational transformation that now makes what was previously invisible visible, nevertheless now makes repeatable and intersubjective the newly discovered moon mountains. In these initial ways, the Galilean telescope is an analogue to linguistic writing, a technology that makes meaning-structures reactivatable, repeatable, and intersubjective.

And although this is probably all that we can legitimately draw from Husserl, the opening of a notion that instruments, technologies, can serve as mediators of meaning-structures can be extended to whole groups of technologies. Recording devices can do for speech what writing does for textual language in that the playback can be repeated and is intersubjectively available for reactivation at any time. Similarly, photography combined with telescopy can make moon images also repeatable and intersubjectively available, and so on. In short, there are many ways in which meaning-structures may be made available in the Husserlian sense, now extended to technologies.

But there is also a way in which Galileo's telescope exceeds writing. Writing is a double transformation—it transforms verbal sensory presentation into a visual presentation, but it also calls for special "hermeneutics" if it is to be read. Whatever referentiality it has must be learned in terms of a set of language skills, and thus the referentiality to a thing itself is not perceptually transparent. The text may be presented passively, as Husserl says, but it also must be reactivated, which brings into play the reading skills of the perceiver. The telescope, in contrast, while it transforms the "thing itself," the mountains of the moon, does so by way of an isomorphic magnification, and the instrumentally perceived object retains perceptual transparency. Galileo with a telescope is considerably more than a calculator or mathematician. Galileo with a telescope is also a perceiver and a practitioner within a now technologically mediated, enhanced world.

And it is here that we need to return to the Husserlian claims that Galileo's mathematization process produces a "forgetfulness" of the bodily-perceptual lifeworld. My claim is that in practice regardless of rhetoric or publication, Galileo never leaves the lifeworld. It is what he sees, albeit mediated by the telescope, the full plenary richness of the Jupiterian "stars" or the spots of the sun, which changes how he understands the newly opening universe of meaning-structures not available to Aristotle, the Church Fathers or the biblical editors. It is just that the telescope precisely, through its transformation of perception, makes dimensions of the newly enhanced lifeworld open to perceptual-bodily experience.

Why Forgetfulness? And Whose Forgetfulness?

The science/lifeworld distance that Husserl claims originates in Galileo is admittedly enhanced by some of Galileo's own rhetoric but not by his praxis. However, Husserl is also "forgetful" insofar as he ignores the transformational mediation of the telescope within Galileo's praxis.

Galileo explicitly made two seemingly contrary claims about the telescope: first, he claimed that anyone could see the Jupiterian "stars" or the mountains on the moon and thereby see more and better than the ancestors. And he explicitly claimed that telescopic mediated sight was better than unaided eyeball sight. But he also claimed that this everyman attainment was only possible under instruction, that is, Galileo's.[12] This seeming contrariety dissolves under a quick and easy phenomenological analysis:

Galileo takes up his telescope, aiming it at the moon. The quasi-transparent instrument, now literally in "mediating position" between Galileo and the heavenly object, produces a profound set of transformations. These include the magnification of the moon such that for the first time details of mountains, seas, and craters immediately are visible. This magnification, however, is not without cost because the moon thus made visible now ceases to be placed in its normal, expansive location within the vault of the heavens. In short, what we today call apparent distance is a phenomenological result of telescope use. Were Husserl to have analyzed this apparent distance with the aid of his own notions of intentionality, he would have noted that the moon as noematic correlate in relation to the observer's bodily position or noetic correlate has reduced the spatial distance. And it is phenomenologically irrelevant if this is described as the "moon is closer to me" or "I am closer to the moon." Nor does this exhaust the instrumental transformation of experienced spatiality. Just as the mountains on the moon are suddenly "closer" within telescopic magnification, so also are my own bodily motions magnified in the use of the telescope. (The user's hand holding ancient telescopes have to learn how to "fix" the moon, and that is part of Galileo's instruction. A tripod helps, but that magnifies the apparent speed of moon motion, and one has to constantly adjust the telescope to the moving location of the moon. In all of this, there are lessons and knowledges that were learned in early modernity, but all because of the mediational experiences of science/lifeworld praxes involving the technologies of mediation.)

We can even, in retrospect, now see how Galileo's telescopic praxis was itself selective. Galileo was much more interested in what lies "out there" in the motions of heavenly bodies and astronomy than he was with his own bodily self-knowledge, which I have hinted was equally made present by telescopic praxis. It was only much later that the reflexive instrumental knowledges became more interesting; yet today in simulation devices, with the gradually growing sophistication of virtual reality processes, it is largely this bodily self-knowledge that is being enhanced.

The Perceptuality of "Theory"

If Galileo's telescope mediated science and the lifeworld, thus leaving Galileo in a lifeworld to begin with, there is one final step that I should like to take. Both Martin Heidegger and Thomas Kuhn have noted and commented on how differently Aristotle and Galileo "saw" motion. For Aristotle, the swaying chandelier in the Pisa Cathedral was a case of "restrained fall," but for Galileo it was a "pendulum." Both Heidegger and Kuhn recognize that in this there is a serious gestalt difference related to a complex background that is incommensurate between Aristotle and Galileo. And whereas Heidegger sees this as a disruption in the history of metaphysics and Kuhn sees it as a Wittgensteinian "duck–rabbit" gestalt shift, both are instances of how a thing is seen.

But this seeing is cultural and hermeneutic, not simply bodily-perceptual. Yet this is the case as well for all seeing—without bodily and sensory sight there is no sight at all, but there is not mere seeing without its meaning-context within which it fits. Husserl's lifeworld does contain both of these dimensions, the bodily-sensory basis that falls under his descriptions of plenary things and the history of praxis that falls under the notion of "origins." That he failed to take the synthesizing move which takes both as "perceptual" also plays a role in his forgetfulness of Galileo's telescope.

3

Embodiment in Reading-Writing Technologies

This chapter describes a transition from Husserl's classical phenomenology to postphenomenology. Here I examine, primarily, Husserl's own technologies—reading and writing technologies, including his eyeglasses for his aging sight problems, his own writing style, which uses pens and stand-up and then sit-down desks, the role of the typewriter and its associated copy processes with references to early twentieth-century scholarly distribution—and then I contrast these with his informal comments concerning scientific technologies, those of optics, and other tools.

I return first to a theme introduced in Chapter 1, the finite life-spans of technologies. Because I am going to discuss the early twentieth-century technologies familiar to Husserl there will be a generational gap here for contemporary readers. As with the Acheulean hand axe, many of the technologies used by and familiar to Husserl are simply no longer used today! Indeed, many younger graduate students who, already in their youth, began with iPads, laptops, word processing, and contemporary electronic means of communication will see typewriters, carbon paper, and "snail mail" as but interesting historical artifacts—let alone quill and dip pens and inkwells. The phenomenon of historical foreshortening is also well known. For some, these recently obsolete technologies may not be that distant from hand axes in youthful historical memory—out of use, out of mind. (My older Husserl scholar colleagues will, of course, well remember what I am about to describe. I include my own anecdotal experience in this analysis.) This generational difference is yet another variant on historical heteronomy.

To begin, as remarked on in Chapter 1, a most important writing technology of the late nineteenth century was the typewriter. These, at first mechanical, writing technologies began to appear in the 1860s—the first American patent was to Charles Scholes in 1867—but of interest here were the first writers to be attracted to this machine. The Scholes machine was taken over by Remington, an early American typewriter manufacturer. Mark Twain and Friedrich Nietzsche were among the first to use typewriters for composition. Twain himself bragged about being the first literary writer with a typewriter—he was a Remington 2 user. Nietzsche's was a Malling-Hansen writing ball, invented by a Danish clergyman, patented in 1872, produced in 1878 (Nietzsche got his model in 1882). It was lightweight and portable with the keys in a ball shape on top which mechanically impressed the letters onto a curved platen for the paper underneath. Of special interest, the intent of Malling-Hansen's designer was to invent a machine that could help myopic or even blind people write—Nietzsche was myopic and loved the machine which he thought particularly good for writing poetry. He wrote at least sixty of his works on this machine. Friedrich Kittler, the cultural historian who has written extensively on typewriter history, pointed out that one "unintended side effect" of the invention of the typewriter was the social change in the secretariat. Most secretaries prior to the typewriter were pen-using males. Many, on the introduction of the typewriter were "Heidegger-like Luddites" and refused to use the machine. However, as Kittler points out, young middle-class women seeking to escape the restrictions of domesticity, and equipped with a similar pre-embodiment skill—most played the piano—readily adapted to the simpler keyboard of the typewriter, and so within two decades most secretaries were female.[1] Writers, including Twain and Nietzsche, early were joined in the twentieth century by many name-visible writers who normally composed on the typewriter: Hemingway, Faulkner, Isaac Singer, Herman Hesse. Not Heidegger, who is well known for his disdain for the typewriter as an inauthentic mode of writing.

This same technological change and abandonment actually preceded the typewriter in the history of pen technologies. As the timeline to follow shows, quill pens were dominant prior to the invention of metal nibs. And although the embodiment skills of quill to metal nib were of minimal and subtle difference, the metal nib pen did eliminate the frequent re-shaping and sharpening needed by quills, and as with the typewriter, within about two decades nib pens had mostly replaced quills.

Do not forget that Husserl's birth in 1859 preceded the invention of the typewriter. Thus it was a machine that was new in his youth. Husserl himself apparently appreciated, but did not compose with, a typewriter. As noted, he was claimed to write standing up (claimed by Herbert Spiegelberg and Karl Schuhmann). However, both Rudolph Bernet and Thomas Vongehr of the Husserl archives claimed he wrote sitting down. Vongehr tells me, "I am pretty sure that Husserl preferred to write in a sitting position at a desk. In

Leuven we still have Husserl's desk which he used nearly his whole academic career. Several photos from different periods show the sitting and working Husserl at this desk."[2] Vongehr goes on to point out that one of Husserl's older philosopher friends, Thomas Masaryk, a fellow Czech, gave him a stand-up desk, which Husserl then gave (in 1934) to yet another Czech philosopher, Jan Patočka. So at one point there were two desks; the latter one has been lost. The same ambiguity adheres to Husserl's writing tools. As Vongehr points out, "As you know, Husserl was working with his manuscripts a lot: reading, underlining, writing comments at the margins, putting the manuscripts in envelopes and writing summaries or short titles on the envelope-pages."[3] His instruments were both pens and pencils—the latter often multicolored. Husserl himself mentions these in his *Transcendental Logic:*

> I have in consciousness a pencil, pen and paper, but no relationality. Or I may begin with considering the pen as *in* the inkpot and as *on* the table that is a relational perception . . . . [or] "thicker" "slimmer" [script (?) lines] are apprehended.[4]

Here, once again, we meet with Husserl's vestigial Cartesianism. Objects with predicates appear first, to which are added relations—no sense of use, no sense of action, no sense of "withdrawal." These are pen-things that are looked at. So we have desks, instruments in consciousness—and we have manuscripts! Although one must rely first on what I call *material hermeneutics,* letting the materials "speak." An examination of the written pages shows a writing style. Vongehr came to the same conclusion I make from embodiment practices. Dip pens call for only a few phrases before the next dip, and both quill and nib pens allow for the script to vary from thin to thick lines. Vongehr notes, "Husserl used most of the times a feather/spring with an ink-pot (this is also related to the fact that ink is the best way to practice the shorthand which Husserl used)."[5] Thus while Husserl's own quotation describes a "pen-thing," Vongehr and I are reading practices into this. What this shows is that for much of his career, Husserl did use a dip pen. Later, the manuscripts show a narrower and more uniform style of penmanship, which, also evidenced by photographs, indicates he switched to fountain pens. (See Table 1.)

I now draw from my time, which, ironically, includes an important regional closure of typewriter technologies. In 1985 I became Dean of Humanities and Arts at Stony Brook—I had already myself begun the turn to computers and my first was a 1984 IBM PC. For those who remember, this machine called for two five-inch floppies to boot up and was very clumsy as a word-processing device. I also had in the same time (1986), an old Zenith portable of the same vintage. (Zenith began doing portables in 1979 and was early associated with Microsoft.) Both such machines are now obsolete and listed in obsolescent computer museums. Moving to the dean's office coincided with a SUNY decision to "computerize" all administrative offices. Ironically, prior to this time our acquisitions office would not let departments buy word-processing computers; after this time they would no longer allow typewriter purchases. (Note that by then typewriters were electric, most with exchangeable typing balls that miniaturized the letter-striking machinery, echoing what had been the ball keyboard of the Malling-Hansen.) On moving to the new office, I found we were using Word-Star, IBM's first word processing program, 1979—exceedingly clumsy, calling for three-key strokes for many commands. Science departments did have computers before this time because of their massive need for data, and these were old DEC machines, now also obsolete. This local ending of typewriter

Table 1. Timeline of writing, reading, and optical technologies

	Writing technologies	Reading and optical technologies
425 BCE		Magnifying glass (or fire starting)
500–1900 CE	Quill pens	For reading small print (Roman times)
1200s		Eyeglasses
1590		Microscope, Jannsen Brothers
1609		Telescope, Lippeshey
1610		Telescope, Galileo
1790s	Pencils	
1800s	Metal nib pens and decline of quill pens	Detection of electro-magnetic radiations:
1800		Infrared, Herschel
1801		Ultraviolet, Ritter
1820s	Fountain pens	
1827	Cartridge pens	
1859 Husserl born		
1867	Typewriters: U.S. patent, Scholes (becomes Remington)	
1872	Malling-Hansen writing ball (Denmark)	
1882	Mark Twain composes on Remington typewriter Friedrich Nietzsche composes on Malling-Hansen	
1890s		EMS instruments: Wireless telegraphy
1895		X-rays, Roentgen
1920		EMS instruments: Practical radio
1931		Radio telescope
1938 Husserl dies		
1940s	Ballpoint pens	
1979 (Ihde 1986)	Zenith portable (word processing)	
1983 (Ihde 1984)	IBM PC (word processing)	
1985	SUNY stops buying typewriters Internet age	

history was roughly a century and a quarter after its beginnings—1860s to 1985. My personal history with typewriters went back to the "touch typing" class we all were required to take in high school (1950s) and what turned out to be, later, a very valuable bodily skill. On to mechanical typewriters—my first book was composed with an Olivetti 32 portable, a fully mechanical machine. Later I had a series of eventually electric and, for a few years, a battery-powered typewriter. Today only the battery-powered one still exists, stored under the eaves in my Vermont house, no longer used. Before discussing embodiment more thoroughly, note that each type of machine called for nuanced different bodily skills. Mechanical typewriters call for a much stronger strike pressure than electrics and are slower. Later, electrics began to introduce technological refinements such as erasure capacities and sometimes time delay and rewrite capacities before word processors. Word processors require only light touch, and some have virtual keyboards. One experiential factor is also that once a subsequent skill is learned, it is very hard to readapt to earlier machines—I doubt that today I could effectively use my first mechanical portable! The year, 2014, marked the closure and death of the owner of the last mechanical typewriter store in New York City—admittedly already a kind of antiquity. Today "antique" typewriters and also radios bring high prices—including Cormac McCarthy's Olivetti 32 (identical to mine), which was totally worn out, but still sold through Christie's for $254,500. McCarthy claims to have written five million words on it! It was replaced by a friend with another, barely used, Olivetti 32.

Return to Husserl in the early twentieth century. Although he did not compose with a typewriter, this machine was familiar and played a central role in the publication and distribution of writing, including Husserl's phenomenological writing. Manuscripts had to be typed for publishers (and later Husserl's shorthand had to be transcribed and typed). Interestingly, apparently the writing "tool kit ensemble" of Husserl's group did not include another machine, which had been invented not long after the typewriter—the spirit duplicator, later commonly called the "mimeograph." These copy-machines, which used stencils, began to appear in the very late 1890s but apparently were not common until around 1918. Rather, according to the Husserl *Briefwechsel* volumes, Husserl and his group used typed carbon copies. The paper for carbon copies was often translucent, almost transparent, thin and could only be produced with limited copies on a mechanical typewriter. (Note: My own MDiv thesis on Berdyaev [1959] exists in this form in my home archives! The original typed copy went to the theological school library.) Betsy Behnke points out that in an exchange with Landgrebe, Husserl mentions that his eyes are weak, "So please don't send the lightest carbon copy of the transcribed mss." He also asks for him not to use transparent paper, to which Landgrebe suggests he put a white piece of paper behind it to better read it.[6] In short, we are here looking at the constraints of that era's writing technologies. Behnke points out that there was a lot of correspondence about numbers of copies and who should get what.

The constraints of the writing technologies, which stimulate the complaints, are matched in Husserl's case by his weakening vision. He does recognize that much of this is because of aging. He discusses "normal" vision but suggests that as one ages, seeing less clearly becomes the new "normal."[7] This is indeed the case, and Husserl here is appealing to what is common *physiological* knowledge. I parallel this with my own experience: Husserl died when I was four years old (1938) and he was then seventy-nine. In 2014 I am eighty, and although I still do not need to wear eyeglasses for most uses, I began using reading glasses in my sixties. And today a new aging normal has set in. We recently traded cars and our new one comes with a complex "control panel" as the maker calls it. It has a multipurposed screen in the center of the control panel (GPS, audio system, rear camera, etc.), which, when turned on first displays a warning: "Do not be distracted by the control panel." Below is a complex set of small buttons with icons for climate control, lighting, and such. I have just learned that by eighty most adult vision includes more difficulty in seeing small, discrete letters or icons while the rest of the visual field remains "normal." One cannot so quickly change focus back and forth. This, however, is somewhat of a special case, since early, as I learned to embody the control panel, the visual switching is a reduced-to-vision embodiment. But I am also confident that this is as much a matter of learning a more whole body embodiment. Until I learn kinesthetically—with tactility and bodily motion—where each button is, the reduced-to-vision-alone process remains clumsy and not fully embodied. It is a sort of keyboard analogue—to look for keys is clumsy, but once one "touch types," one no longer looks for keys at all. This is actually a phenomenological-postphenomenological point. What is primary is whole body action and motility; focused reductions to vision or any other sensory dimension have some reductive results. And, indeed, on our most recent and only third trip, I have begun to be able to kinesthetically locate many of the buttons now learned and embodied.

Return to Husserl: In his own case, his failing eyesight cannot be remedied fully by eyeglasses—and his writing about this seems to be lacking any phenomenology of embodiment. So he turns to a magnifying glass! He refers to this both in his correspondence with Gustav Albrecht and in what Behnke calls a thought experiment:

So, in my room we suppose there an invisible ghost, also untouchable, etc. As often as I call upon him, he is "there." I cannot carry a heavy box, he helps me, and we push it and lift it together, etc. There is something I cannot see very well, with my weak eyes I cannot read the small letters, and he reads aloud for me. Afterwards I convince myself with a magnifying glass that he has really helped me. I ask him to bring me something from the street, and he brings it to me, etc.[8]

I must admit when I first read this I was baffled. It comes from Husserl's work on intersubjectivity and the ghost is a sort of "other" subject. But then when comparing this text with what Husserl has to say about bodies, I realized that such a ghost is compatible with his still "Cartesian" body.

As a worldly self, I have shaped my body as body. I have come to know its organs as controllable and I have learned this control in a general sense; it is now my organ—for my life in and rule of the world. . . . In my world-life I learn through it to practically rule nature and bringing the body into play in the different directions of this striving and effect. I shape it, i.e., in particular ways—as carpenter, locksmith. etc. I learn the hand grips, i.e., the hand refines itself in certain directions as an organ and accordingly the entire body refines itself in different activities.[9]

Here is a prime example of how Husserl's Cartesian language betrays any phenomenology; this does not sound anything like "I am my body." Rather, it sounds like a conscious and executive ego, even one sounding like it is outside a body, is shaping and commanding a body—which sometimes he also calls a "body-thing." One can also see that this language corresponds to that of eyeglass-things, telescope-things, microscope-things because the Cartesianism of the day takes physical things as the foundations for perception. In correspondence with other Husserl scholars and secondary sources, there is a consensus that early Husserl remained most closely tied to what Behnke calls the received division of the sciences of the day, "This shaped his strata ontology— the lowest being *Ding*, sheer physical materiality, then built on that *Leib/Seele*, then built on that *Geist* . . . and in more texts the simplification of the scheme into *Natur/Geist*."[10] This almost perfectly follows the notion of modernism as described in the Introduction in my discussion of Latour and Forman; it also confirms Husserl's own modernism.

I will leave this topic with one final example, a tool example of a scissors as described by Husserl:

> The human being there, what he does and what he makes. He rules the body, perceiving, he moves the eyes, fixates on this over there, then on this here, moves his hands probingly. However, the human being has an intention within the field of his own experience. He grasps an object (a piece of paper), holds it in the one hand; with the other hand he holds a pair of scissors and cuts off one strip, etc. That I understand without further ado and individual-typicality. He behaves in the typical way of "a" person, who cuts with "a" pair of scissors."[11]

To me this almost sounds like a description of a robot! But again here is an executive consciousness commanding a "body," which seems passive and effectively machine-like. More interestingly from a postphenomenological perspective, however, is the way in which the action is described as *typical*. Husserl is interested in the typical, the general, and this insofar as these features can lead to what he takes as higher ideality. Even after he himself turns to a more praxis-oriented stance in the "Origin of Geometry," where the emphasis is upon "standard measures," "empirical basic shapes," and in the "univocal" practices, which he then sees can lead to more purified, more perfect, refined shapes as in classical geometry—circles, squares, triangles, cones.[12]

Rather, let us look at what I call a *virtuoso practice* as perhaps in an atypical trajectory. We recently attended the spectacular exhibit of the late works of Matisse in the Museum of Modern Art in Manhattan titled "The Cut-Outs." In late life Matisse was no longer able to stand up and move about doing the large paintings he is so well known for; rather he was mostly confined to a wheelchair and took up more consistently the practice of cutting out his art-shapes with a tailor's scissor! These brightly colored shapes filled gallery after gallery of MoMA, bright, joyful and in the case of the series of "The Blue Nudes" dramatically displaying a curvaceous feminine set of shapes (see Figure 1). Matisse practices a virtuoso cutting skill, and almost as an intuitive phenomenologist, he played with variation after variation, as evidenced by the pin marks on his walls and by the colored paper pieces. This is embodied skill, embodiment in action. This is not "typical" cutting.

Now we can return to Husserl's pens—different pens call for different bodily skills. An ink-and-dip pen is slow and can make very limited numbers of letters before another dip—thus thinking time is greater between dips in such writing than reservoir pens, which can produce much longer sentences and paragraphs between fills. Neither technology is neutral, and although neither can "determine" writing style, each "inclines" toward subtle differences in style. This, too, is part of the embodiment process. Today's word processors favor quite different writing capacities. Moving paragraphs about and, indeed, the entire rewrite and editing process are much easier.

Figure 1. Henri Matisse, *Blue Nude IV*, 1952.
© 2015 Succession H. Matisse / Artists Rights Society (ARS), New York.
Photo: G. Dagli Orti. Musée Matisse, Nice, France.
© DeA Picture Library / Art Resource Society (ARS), New York.

When it comes to scientific instruments it is clear that a kind of learning is possible that differentiates the learned observer from an unlearned one. Husserl observes, here regarding the microscope:

> One cannot come to an understanding with some one who is both unwilling and unable to see. Seeing is not always a simple matter, not even when considering external nature. The biologist using his microscope sees very much more than the porter; he has learned a type of seeing that the latter has not learned. He cannot really demonstrate this seeing to anyone, and whoever has *a priori* arguments that rule out a certain type of seeing will never be able to be convinced by a microscope.[13]

Shades of Galileo's Jesuits! In their reluctance to see through a telescope they often blamed what today we call "instrumental artifacts" in the instrument itself. I have argued that phenomenological seeing, while clearly a bodily perceptual skill, must become skilled to the point of becoming a connoisseur perception. Yet it would be a disaster were Husserl right about not being able to demonstrate or teach such a kind of seeing. The porter nor anyone else could ever become a biologist. Contrarily, people can be taught to do wine tasting, to recognize the multitude of different sparrows among the little brown birds as in bird watching, or to recognize the different kinds of mushrooms that are edible—all are acquired embodiment skills. Yet what the microscope introduces is its magnification, which makes visible the previously too-small-to-see. It is a mediator for new perceptions beyond the ordinary parameters of unaided vision. For example, in the earlier mentioned telescope parties in Vermont, our viewers are apparently willing to learn and to see. Since Jupiter's satellites move so rapidly, one can quickly see that they circle the planet. Moreover, we have now had four centuries of accumulated experience with trustworthy optics. No one to date has attributed even first viewings to "instrumental artifacts."

Husserl's recalcitrant viewer who refuses to see, seems to do so primarily because of conceptual issues. Instrumentally mediated vision, however, entails learned embodiment skills. These can be taught, although it remains the case that only some observers will become virtuoso viewers. Today scores of postphenomenological researchers have performed case studies on such skill acquisition All technologies in use are interrelational and imply embodiment skills. Each difference interrelates with different, but learnable, skills. Let us, as a variation, return to the philosopher's study today in contrast to Husserl's early twentieth-century setting. First, the writing technology: Both Husserl and Heidegger composed with pens, Husserl standing and/or sitting, Heidegger seated, at their desks. With the introduction of the typewriter, as indicated, Nietzsche already at the end of the nineteenth century, followed by many—if not most—serious writers had changed to typing for composition. In passing, both he and Mark Twain had moments of doubt relating to typing. Twain thought it corrupted his morals and Nietzsche's Writing Ball was damaged in travels, and since this was before typewriter repair people existed, his mechanic—unskilled— actually made the situation worse.

Jumping to my own end-of-typing era, I note that by the end of my term as dean, a poll showed that by 1990, 85 percent of the English faculty used word processors to compose. And although today a few faculty still compose with pen and yellow, lined pads, I doubt that graduate students get far from their iPads or laptops. However, if word processing is the composition technology of choice today, the laptop is part of a much more complex technology network: printers, now virtually all multipurposed, do printing, copying, faxing, scanning, and are tied with the laptop into the internet, which can quickly transmit anything which has been composed to anything downloaded from the "net or other sources. This technology network is so radically different from that of Husserl's time—and it can even accommodate failing eyesight. Enlarge the font; raise contrast with the bold key; make copies that are clearer than any carbon copy. In this era, a copy is frequently *better* than an original. This technology change is less than a century since the 1930s, when we first took note of Husserl's reading-writing technologies. Before looking at more embodiment relations, take note that this contemporary reading-writing complex also has vast implications for the production and distribution of philosophy and other academic discipline output. Both dissemination and intake have been made fast and easy. If one takes a look at the typical study of an academic, one is likely to find one or more computers, laptop and desk computer, iPads, multipurpose printers, reading technologies such as Kindle, cell phones of various kinds, possibly land lines, plus equally complex "entertainment" technologies sometimes in special entertainment rooms—both audio and visual. This saturation has both intake and output capacity. Any smart phone connects virtually to a universe in information-producing sources

Philosophy distribution is a special case. Historically, if we go back to our Greek origins, writing remains are rare but exist as fragments (Pre-Socratics), dialogues (Plato), books (Aristotle's remaining ones far outnumber Plato's); by early modernity and early science rapid delivery of short book-length works began to occur. Copernicus was distributed throughout Europe quite rapidly, and by late modernity Roentgen's X-ray photo of his wife's hand with ring was made into a postcard and distributed within weeks.

Now, however, one attaches an e-file to an e-mail and sends with the speed of light! (Indeed, several sources for this book arrived in just that way to my Manhattan study. In addition, of course, were my many viewings of information servers on another multipurpose variant of my word processor-computer-screen.)

This calls for a few brief descriptive references to perhaps the now dominant embodiment practices for contemporaries. We spend lots of time seated in front of screens (computer, television, mobile screens, and in my apartment high-rise our gym machines are connected to screens that we can watch while exercising). Screen ubiquity is too familiar, and as a philosopher of technology, I have often been engaged in critical and reflective projects related to contemporary worries about this set of embodiment practices:

> Are these practices too reductive? Gaming, in many of its forms, reduces motility to focal eye-hand coordination movement. Does this enable part of the trend to obesity? Yet it also hones skills for what today is called "Nintendo surgery," the kind of laparoscopic and robot surgery so pervasive today, which has replaced the massive open surgeries of the past.
> Can new forms of technologies change these problems? Wii games engage more bodily action, and virtual reality simulation technologies entice far more holistic experience than sit-down gaming. Phenomenologically, of course, all these variations also transform our experiences of space-time. Simulators today are also related to distance sensing, robotics and drones, and the imaging breakthroughs into previously closed or unknown forms of reality.

Postphenomenologists are delving into precisely such phenomena. The first special issue journal to account for postphenomenology research was the 2008 issue of *Human Studies*, vol. 31, no. 1, published a year following our presentations as research panels at various STS conferences. All four research articles in that issue are typical examples: Peter-Paul Verbeek on ultrasound and how it transforms the experience of a fetus; cell phones and the Grameen Bank experiments in empowering phone ladies by Evan Selinger; Cathrine Hasse's study on learning to read science images; and Robert Rosenberger's study of embodiment and the Mars Explorer. Add Anette Forss from the Karolinska Institute and her study of how pap smear readers learn to see cancerous formations, and I could add my work in *Bodies in Technology* (2002) and *Embodied Technics* (2010), which deals with cloud computing, robots, virtual reality, and so on, as other examples. Although this is schematic, and although I will not here go into extended detail, I am showing that with each new technology there are interrelated embodiment skills that implicate us embodied humans and for which there is a possible postphenomenological analysis.[14]

4

Whole Earth Measurements Revisited

The Society for Philosophy and Technology met in Puebla, Mexico, 1996, and I presented a paper titled "Whole Earth Measurements," a bit tongue-in-cheek, which looked at the problem of global warming. My subtitle was "How Many Phenomenologists Does It Take to Detect a Greenhouse Effect?" and I turned to both Heidegger and Husserl as my exemplary phenomenologists. This was before today's grand political debate about climate change with so many right-wing people not only denying global warming but arguing particularly that the homogenic factors claimed are a great hoax. Here I return to that issue as a crucial test case for missing technologies—scientific instruments—and classical phenomenology in Husserl's form.

This revisitation will be unapologetically "postmodern" in Paul Forman's sense. Science here will be, strictly, *techno-science,* that is, science as fully embodied in its technologies, many of them new to the twentieth and twenty-first centuries.

In this sense, it will be taken for granted that not only does science needs instruments both for measurements and for observation but also for the very discovery or constitution of new phenomena. Returning to themes from the preceding chapters, without the telescope none of the phenomena that constitutes the solar system would have been known, and without the microscope the microworlds that eventually led to modern biology would not be known either. Thus in the Forman sense, this is postmodern.

Then, too, global phenomena, and in particular global environment issues, are also a new or a contemporary set of problems. Although many ethnologists have argued that certain traditional societies were environmentally sensitive, seeking implicitly to have a sustainable nature-culture (I myself have noted inland Aborigines in Australia, Inuit in the Arctic), many other traditional societies are implicated in massive extinctions (wingless birds in south Pacific cultures, megafauna among the early Native Americans, etc.), none had a global sense of environment. Global environment is a new problem.

> Some technoscience preliminary factors must be noted:
> First, the issue of climate change is a whole earth phenomenon. To detect it one must measure an earth globally, not just some sample set. Thus in contrast to much earlier science laboratory experimentation that narrowed and constrained a field, whole earth measurements must be global. The whole earth becomes a very different "laboratory."
> Climate phenomena are extraordinarily complex. Atmospheric and oceanic phenomena, the ice-melt of glaciers, volcanic phenomena, industrial activities, transportation, and even dung and charcoal cooking— all enter the mix. Such a science is ingenious in finding

substitute measurement indicators. Ice cores, for example, go back 400,000 years in Greenland, but gas bubbles yield ions for mass spectroscopic analysis to determine atmospheric levels of greenhouse gases. Not only must measurement observations take many forms, but the role of simulations, which are increasingly used for complex phenomena, enter the scene. Here I note that Monte Carlo or multiple simulations have been used since the Manhattan Project, to a now ordinary use in projecting pension distributions. (When I retired in 2012, my pension adviser ran over five hundred variant simulations to cover the multitude of eventualities. For example, actuarial life expectancies have gone up drastically such that while no one can predict how long a single individual may live, a set of probabilities are simulated to determine safe outlays to cover the most probable outcomes.) The same process is used for climate projections.

Today's sciences are probabilistic. Indeed the "Daubert Decisions" of the mid-'90s changed the legal definitions of what counts for scientific expertise from a general sense of consensus to the 95 percent probability that today is regarded as reliability in science. In passing, I would point out that the consensus of scientists regarding global warming in 2014 with homogenically significant causes was shown to be a 97.2 percent agreement evidenced by seven surveys done that year.[1]

Can Classical Phenomenology—Husserl—Detect a "Greenhouse Effect"?

In my original article I phrased this question more broadly, but here the presupposed context is the Husserlian version of classical phenomenology with specific reference to what I am calling his missing technologies. As I have suggested, whole earth measurements are different, more complex, and in phenomenological terms require a new perspective for measurements. Even if one were to take a first *synchronic* global measurement, this would be difficult but insufficient. For example, one periodic and significant causal factor in the climate system is the El Nino cycle, a periodic multiyear hot spot near New Guinea that heats up the ocean and the atmosphere and affects ocean currents. But cold spots elsewhere might counterbalance this phenomenon on a global scale. So thermometers must be scattered globally, surface, subsurface ocean, land, and so on, and connected to computer tomography for overall results. Then *diachronically,* one must consider that this might be a few-year blip, so very longtime records must come into play; however, the complication arises that actual past temperature measurements are but a few centuries old. The passion for measurement is itself modern. So as noted above, longtime measurements of ice cores, tree rings, sediment layers, and the like must come into play. Today, everyone knows the result: the "hockey-stick" upturn shown on charts of global temperature in the last century over a record now at least 400,000 years old. In addition, what is unique about this record is the politically sensitive conclusion that much of the hockey-stick phenomenon relates to industrial and other homogenic factors that began with the Industrial Revolution and continue today. As early as 1994 *Science* magazine estimated that 24 percent of greenhouse gases were caused by industrial activity.

But how does this specifically relate to phenomenology? My claim is this: I hold that a unique *perspective*, which I will call earth-as-planet, and an understanding of measurement practice from a thorough *technoscience,* or instrumentally embodied science, are called for. Classical phenomenology has neither of these concepts. As noted in the introduction, Husserl remains a Forman-like modernist in simply subsuming instrumental materiality under science. For Husserl, at base our knowledge is constituted by way of our perceiving bodies with perception of sensory plena, whereas instrumental mediation for Husserl yields a perceptual-correlate—although it may improve earlier forms of knowledge. Let us take as a specific example greenhouse gases (CO_2, CFOs, ozone gases produced by industrial or other human activity—indeed we can include carbon particulates from dung cooking fires!). All these gases and particulates are *subperceptual!* They cannot be perceived in Husserl's primary dator perception. Rather, in Cartesian ways they can be inferred. But, I claim, if science is technoscience and instrumentally embodied, such particles and gases become perceivable in visualizable form through imaging technologies. This, however, calls for a phenomenology of instrument use which recognizes seeing with and through instruments.

At the heart of Husserl's version is a strong distinction between primarily perceived plena, which are directly sensorily perceived, and mathematized objects (the pure shapes of geometry), which are both abstract and idealized and only indirectly available, "[Science in] . . . all this *pure* mathematics has to do with bodies and the bodily world only through an abstraction, i.e., it has to do only with *abstract shapes* within space-time, and with these, furthermore, as purely 'ideal' limit shapes. *Concretely*, however, the actual and possible empirical shapes are given, at first, in empirical sense-intuition, merely as 'forms' of a 'matter' of a sensible plenum: thus they are given together with what shows itself, with its own gradations, in the so-called 'specific' sense qualities: color, sound, smell, and the like."[2]

Husserl thus has a conceptual duality between concretely perceived plena and abstractly idealized pure shapes. (How

Cartesian can you get?) Moreover, what he calls an indirect process is where science in practice uses and places instruments. "Now with regard to the 'indirect' mathematization of that aspect of world which in itself has no mathematizable world form: such mathematization is thinkable only in the sense that the specifically sensible qualities (plena) that can be experienced in the intuited bodies are closely related in a quite peculiar and *regulated* way with the shapes that belong essentially to them."[3] Again, this sounds Cartesian with extension and shape to *res extensa*.

Return to the greenhouse problem: Were Husserl right, then, the (unperceived) gases are constituted through idealization and mathematization indirectly. What is wrong relates to the materiality of the gases and the instrumentally mediated perceivability of the gases themselves. Instruments, I argue, maintain the connection of the sciences to the lifeworld. The problem, however, lies with Husserl's flawed notion of science. It is science that has undergone an abstraction, since in practice it is embedded materially in technologies—instruments. Instruments provide both measurement and perceptions that relate to the "things themselves." Actual science is *embodied through instrumental extension of primary perception through instrumentation.* Put simply, CFGs, CO_2, and ozone are not pure shapes but are materially presentable material entities instrumentally mediated. Science is not disembodied but instrumentally embodied in an instrumental realism—if it is a technoscience.

Whole Earth Measurements Redux

Return to the original problem: How can one tell anything about a greenhouse effect phenomenologically? I claimed that to do so one must use as a frame a notion of science as embodied technoscience and one must have a perspective on earth-as-planet. The first is what provides a referential realism regarding the phenomenon. Measurements must be of something, and what instruments do is precisely open the world to its micro- and macrofeatures. In early modernity this was dominantly through optics (microscopes and telescopes) but in postmodernity through the vast array of imaging technologies, which include mass spectroscopes, scanning tunneling microscopes, Hubble orbiting telescopes, and more. Through this panoply greenhouse gases are more than inferred; they are instrumentally perceived.

Perspectivally, the very notion of a whole earth perceived is best noted in the perception of earth-as-planet. The earth as circular, finite, and material, is, of course, ancient. In Aristarchus to Copernicus, that is, from Greek to early modern times, the circular earth was depicted. But in another sense, one could note that it was not "seen" as such until orbiting astronauts saw it, earth was photographed from the moon, and satellite imaging made earth-as-planet a commonplace.

Science Praxis Seen as a Postphenomenological Hermeneutic

As noted, contemporary—postmodern—science does its work by means of its technological embodiment in instruments. Today's array is vast, running from massive particle colliders such as CERN, down to nano-sized laser tweezers for manipulating single atoms and molecules, or in other forms single photons. Similarly, the array used in detecting climate change is vast and diverse, from satellites to deep-sea-diving submarines. The production of Big Data is immense and runs through complicated computer tomography programs. In many years of research on imaging technologies, I have noted that some imaging is isomorphic, such that the image is in some degree "like" the object imaged.

Although no images are perfect replicas (only in science fiction does this occur), recognizability is clear. But other imaging is nonisomorphic, such that the image does not resemble its referent. Examples would be spectroscopes that produce "bar code-like" images that present phenomena as chemical signatures. (Colored bands of yellow into orange are sodium spectra which show up from the sun to sodium gas.) The contemporary tomography that can convert data into image and image into data is a coded quantification process that I term "hermeneutic." Better, it is a *material hermeneutic* since it constitutes meanings from material. If we take these praxical capacities and apply them to earth science, then as one set of isomorphic imaging processes, imagine satellite, drone, or balloon imaging of earth surface phenomena. Deforestation and sea-level rises are quite clear and dramatically displayed. The melting of glaciers is neither indirect nor abstract but visualized in a gestalt of time lapse. Then if one turns to more specific art-like variations— the use of infrared or ultraviolet photography—even greater contrasts emerge, again as time-lapse gestalts. Carried over into public communications, who has not seen videos of tsunami floods? Calving icebergs? Raging forest fires? In my own case, the closest to a vision of Hades I can remember is the out-the-window vision of over a thousand miles of oil field smoke plumes flying over the devastation of the Kuwait War oil field arson while on the way back to Australia from Italy in 1991.

I now want to draw this chapter to its close by taking note of a few significant ways in which a postphenomenological set of practices works within science praxis. First there are the practices that use instrumental phenomenological variations. Husserl made "imaginative variations" a core practice within classical phenomenology. By varying a phenomenon he claimed to find "essences" or structures. Instrumental variations, that is, variations with different instruments, attain similar insights into "things." For example, if one wants to find the date of some thing, then the use of carbon 14, thermoluminescence, and archaeomagnetism, when calibrated to the same range can converge and thus produce a robust "fact" or date. Or if turning to imaging, the use of optical imaging can have infrared imaging added to make for better contrasts between organic and inorganic things. Add more, such as thermal imaging or magnetometry, and one gets even more precise contrasts and even subsurface features. In short, phenomenological variations embodied in instrumental investigations map science and phenomenological praxis on each other.

Second, Husserl's use of variations aimed at producing *invariants,* or essences. Postphenomenology—using variational methods—often finds *multistabilities* instead. These are multiple ways of seeing, of multiple arrangements, and variants on themes. In recent decades many of the sciences have also discovered this complex phenomenon and have dealt with multistability. For example, in surface microchemistry thin films (two-atom thick films, for instance) can self-organize in multiple (5 to 13) ways. My former colleague and university president discovered multistability in optics. And today's astronomy has a large number of galactic stabilities out of the billions now charted.

Hermeneutics or critical interpretation is also crucial to science praxis. For example, to "read" a material "calendar" of the past, comparing tree ring to sediment layers to patterns in ice cores can show when and for how long there were wet compared to drought periods. Such a hermeneutics of things (not merely texts and linguistic phenomena) are part of a postphenomenology as well.

Third, a postmodern science is no longer modern in another sense—it is active, constructive, and reconstructive from its sense of perception on. Instrumental perception is not receptive or passive in the early modern sense—it is actively constructive. This is implied in the variational method of the first science/phenomenology parallel. For the best imaging, the results must be manipulated. One simple example is in the role of "false color" techniques. (I dislike that terminology and prefer either "relative" or "assigned" color.) By varying colors in images, one can sharpen, enhance, and often discover features not obvious or even visible to ordinary optical imaging. Interestingly, such techniques have longer been known and used in art. Another example, previously noted, is the use of multiple simulations to elicit patterns out of simulation variations. In one sense, the more the better.

Climate Changes, Old and New, and the Homogenic

In this chapter I have been looking at the shift from modern to postmodern science and its relation to phenomenology. For a conclusion, however, I want to add some consideration for premodern climate and ecological change. If one takes a long view—for example, looking at the great climate changes and extinction episodes in earth's history—it is obvious that major climate changes and massive extinctions occurred before humans evolved. Thus were one to retrograde today's arguments about homogenic contributions to these events, it would be clear that no homogenic causes were involved.

However, once we arrived, and now drawing from postmodern scientific knowledge, it becomes apparent that even premodern climate and ecological changes had homogenic causes. I noted above that differences in human cultures with respect to sustainability have varied, and thus migrations that included those of early Americans had homogenic factors related to the extinctions of megafauna. Premodern Pacific oceanic migrations led to the extinction of wingless birds. And the extinction of early American fauna—the last deliberate killing of a passenger pigeon—or near extinctions of buffalo are better known. In each of the extinction cases the environmental complex was permanently changed. And in each extinction involving homogenic factors there are also technological dimensions—from stone weaponry for mastodon hunting to rifles for passenger pigeons.

Today's science is not only global, but much more related to system complexity. Thus today one can recognize that top predators have vast "cascade effects" upon ecosystems. The reintroduction of wolves into the western states has ended up restoring even overharvested flora (aspen trees devastated by elk). Recent studies have shown that baleen whales are major factors in keeping ocean health working. They distribute krill over previously decimated areas and thus help sustain feeding areas. Systems and interspecies patterns involve postmodern science's multiple instrumentation tied to computer tomographic capacities. Today's postphenomenology, which takes account of human-technology relations in praxis, can detect climate and environmental changes.

5

Dewey and Husserl:
Consciousness Revisited

Phenomenology as a twentieth-century philosophy has since
its classical beginnings been portrayed as a subjective philosophy and frequently claimed to be antiscience. I argue
that both characterizations are false or distortions of phenomenology, and so a modification of classical phenomenology is needed, and that is postphenomenology. This chapter
follows that trajectory in a somewhat different way. The late
twentieth century began to see two science paradigm shifts in
the Kuhnian sense. The first was a revival and renewal of
interest in consciousness and the other, often closely related,
in animal studies. Here I interweave these two contemporary
movements to a rereading of John Dewey and Edmund
Husserl on consciousness.

The Return of Consciousness

Consciousness has suddenly become fashionable. Let us take note of some of the diverse disciplines and contexts in which consciousness has returned from its mid-twentieth-century situation, hidden in the shadows of behaviorism in psychology and many of the biologically oriented disciplines, and from naturalist reductionisms, again rampant in these same and sociobiological contexts:

> In psychology, consciousness returns with the shift away from dominantly behaviorist to more cognitivist forms of psychology. And although we have heard much about cognitive psychology and its new ways of examining consciousness, we should also note the close association of cognitivism with two other related developments. Both philosophers of mind and phenomenologists have flocked to this movement, but with very different takes (see below).
> Cognitivism today has begun to include many animal studies, thus expanding the notion of consciousness from humans to others within the animal kingdoms. Two examples include the to-be-expected studies with primates, but the less expected discoveries are among birds, particularly jays, ravens, and crows, the corvids. In the primate cases, particularly among chimpanzees, behaviors have begun to be recognized as "cultural" even in contrast to "environmental" behaviors. And from the interests here, one may add technologies. The April 6, 2007, issue of *Science* magazine reports on a conference titled "The Mind of the Chimpanzee" held in Chicago. Note, such a topic title would have been unimaginable a few decades ago! As one participant exclaimed, "We're talking about things now that I couldn't talk

about in the '60s. We couldn't even talk about the chimpanzee mind because chimpanzees didn't have one."[1] Studies showed how chimps could recognize number sequences of 1–9 in sequence no matter where the numbers randomly appeared on a screen. Decoding facial expressions and vocalizations has proceeded to the point that the investigators now know that not only are there warnings about predators but also about what kind of predator is perceived. Different vocalizations are for leopards, snakes, and eagles. And, finally, regarding cultural knowledge—that is, generationally transmitted knowledges as contrasted with occasional innovations in environmental situations—has now moved from the 27 unique cultural behaviors reported in the 1990s to 571 such behaviors today. The jay-raven-crow studies may appear to be even more surprising since some investigators hold that this group of birds may actually equal or top primate cognition. In the April 7, 2007, issue of *Scientific American,* Bernd Heinrich and Thomas Bugnyar, in "Just How Smart Are Ravens?" claim that "recent experiments show that these birds use logic to solve problems and that some of their abilities approach or even surpass those of the great apes."[2] The investigators then report on several such experiments. Ravens, seeing a food piece tied to a string hanging below a branch, survey the scene for several minutes—then at a first try, pull up the string with their beaks, place a foot on the string across the branch and repeat the process until the food item is within reach. In a much more complicated setup two ravens, one able to watch a bird outside the cage cache food, the second not able to see the bird, were then allowed to seek the hidden food. "Knower birds" would then protect the hidden food from the "hider bird," but nonknower birds would not, thus implying that the birds attributed awareness to the hider bird. My own favorite video shows a raven watching an ice fisherman—who leaves his bobber and goes off for a beer. The bobber submerges; the raven flies down and does the bill-and-foot pull, catches the trout, and flies off with it! For our purposes here, note that two things are happening in this investigation: First, and most important, beliefs or prejudices against animal intelligence and by extension animal consciousness have been over-come, and second, experimenters have had to become "smarter" in constructing experiments to show animal intelligence.

I find especially interesting that in many cases, including both primate and corvid intelligence, tool or technol-ogy uses are crucial in the intelligence experiments. Chimps, for example, fashion probes for termite mounds, apparently calculating both the lengths of the probes—making them long enough to avoid the termites close enough to bite—and the number of probes—providing themselves enough probes to guarantee a big enough meal (since the probes wear out quickly). Newer observations show that female chimps, through chewing, construct sharpened spears to probe for bushbabies—a squirrel-like animal—in tree cavities. In the case of New Caledo-nian crows, in their hunt for tree grubs, fashion probes from the stems of leaves, pulling off the green matter and leaving only the sharp hard stem; on being given wires, they innovate by actually shaping hooks to catch the grubs. Thus we are taking animal intelligence both into the realms of calculating consciousness and technological innovation. I shall playfully designate such animals as "Deweyan pragmatists" for reasons to be discussed below.

But the biggest single scientific development driving the revival of interest in consciousness is neurology, and in particular neurological imaging processes. By now everyone is familiar with reports in science magazines, the popular press, and the media, that show "brain scans" with lit-up parts of the brain under different experiential conditions. And with this twenty-first-century "phrenology"—as some, including me, call it—the question is "How does *the brain* decide?" The Cartesian homunculus has now become the object of neurological science, and both philosophy of mind analytic philosophers (the Churchlands, Paul and Patricia, with their "heterophenomenology," and Andy Clark) and phenomenologists (Shaun Gallagher, Donn Welton, and Dan Zahavi) are flocking to the scene. I shall return to these observations about consciousness redux later in this chapter.

Consciousness in Dewey and Husserl

With consciousness again fashionable, I now turn to the two early twentieth-century inventors of experientially based philosophies, John Dewey with pragmatism and Edmund Husserl with phenomenology. Although both made human experience primary for philosophical inquiry, consciousness played radically different roles in these two philosophies. In what follows I show how each of these contemporary philosophers of experience used different models to explicate epistemologies of experience: In Husserl's case the early model frequently was borrowed from Descartes and the famous reduction to a meditating ego also identified with consciousness. In Dewey's case the borrowing was frequently from Darwin and the notion of interaction between an organism and its environment. I argue that in both cases the models call for an interrelational ontology, but the models themselves have different vectors: Husserl's direction makes consciousness, subjectivity, its center of gravity and thus has a transcendental vector; Dewey's makes experiential experimentation its center of gravity and thus takes what is usually called a "naturalizing" vector. I hope to show, however, that this Darwin-inspired ontology is not a reductionistic naturalizing.

I begin with Husserl and his version of an ego or consciousness-centered phenomenology. Clearly, the *Cartesian Meditations* are an excellent example of his early consciousness-centered vocabulary relating to a phenomenological reconstruction of philosophy. Husserl deliberately takes Descartes's *Meditations* as a model for doing a radical phenomenology. In part, this is tactical, relating to the *Paris Lectures* and his French audience. But in so doing, this modeling of a vocabulary of *ego cogito,* subjectivity, and transcendental subjectivity brands this version of phenomenology as belonging to the subject-centered traditions of modernist European transcendental philosophy. Strategically, of course, Husserl uses the Cartesian model to effectively invert Cartesianism: by using "reductions." What this phenomenology does is to transform or translate notions of the subject, the world, and the interrelation between self and world through intentionality.

At this point I must make a meta-observation: Both Dewey and Husserl remain thoroughly modernist. This is to say that the variables reshaped by their respective philosophies of experience remain (a) nature, things, objects; (b) others or society, culture, history; and (c) the largely individual experiencing subject. Linking these three modernist dimensions, in Husserl's case, is intentionality; in Dewey's case the organism/environment interaction is taken in an instrumentalist direction.

Returning to the *Cartesian Meditations,* Husserl, arguing that Descartes's reduction was not radical enough, sets up his series of reductions [epoche, eidetic, and transcendental] to reduce to a descriptive pure conscious experience, whose interrelational structure is intentionality, which remains a structural feature of consciousness in relation to world, others, and even an empirical self, structured overall by inner time consciousness. I paraphrase this strategy as one that first sets up transcendental subjectivity—as consciousness of a special sort, which is then reconstructed by taking world, others, and all experiential phenomena into itself. I shall here give only the briefest exposition of this itinerary from the *Cartesian Meditations*:

> First, the adaptation from Descartes, where Husserl
>   praises Descartes for "a regress to the philosophizing
>   ego . . . the ego as subject of his pure *cogitationes*,"
>   which is then characterized for philosophy thus:
>   "Changing its total style, philosophy takes a radical
>   turn: from naïve Objectivism to transcendental
>   subjectivism."[3]
> Then in Husserl's first meditation, titled "The Way to
>   the Transcendental Ego," he proceeds to set up
>   his transformational rules, the various reductions and
>   their modes of evidence. He maintains the Cartesian
>   flavor by noting, "Evidence is . . . an '*experiencing*' of
>   something that is, and is thus: it is precisely a mental
>   seeing of something itself."[4] Conscious experience
>   remains primordial.
> Phenomenology then refuses to accept anything as
>   evident unless the phenomenon is derived from

"*evidence,* from 'experiences' in which the affairs and affair-complexes . . . are present to me as '*they themselves.*'"5

This applies, first, to the world, but phenomenologically, the world transformed into world-as-phenomenon. The world is thus reconstructed within transcendental subjectivity as an "acceptance-phenomenon."6 Such a world is not nothing, as Husserl maintains, but a world as relational pole to "pure subjective processes ". . . *purely as* meant, with its other pole my self apprehension, 'purely': as Ego, and with my own pure conscious life, in and by which the entire Objective world exists for me."7 World, then, now exists as phenomenon-world. Consciousness has taken it into its pure self-experience.

The same transformation, in the Second Meditation, happens to others or other subjects. Husserl claims that his moves are only at first glance "solipsistic," "whereas the consequential elaboration of this science, in accordance with its own sense, leaves over to a phenomenology of transcendental intersubjectivity and, by means of this, to a universal transcendental philosophy."8 Thus just as world is transformed, so are other subjects transformed as fully intersubjective.

Ultimately, however, the secret to the phenomenological transformations or reconstructions of world and others is what can be anticipated as an implied interrelational ontology under the notion of intentionality. "Conscious processes are also called intentional; but then the word *intentionality* signifies nothing else than this universal fundamental property of consciousness: to be conscious *of* something; as a *cogito,* to bear within itself its *cogitatum.*"9

I shall not go further here since I believe my case is made: Husserl adapts the Cartesian vocabulary of *ego cogito;* adds *cogitatum* to account for interrelationality, but remains under the umbra of "consciousness language" thus leaving his phenomenology open to charges of a philosophical subjectivism. It was this "fatal attraction" to the transcendental tradition, I have held, which negatively cast phenomenology as a subjectivist philosophy for late modernists, including most of the Anglo-American traditions dominant in Anglophone and even some European countries, and even more for the postmodernists who saw the "end of subjectivity" and a deconstructed subject on the Euro-American side.

Now Dewey: I remind all of us that Dewey and Husserl are genuine contemporaries, both born in 1859 and doing their original experience-centered work in the early teens of the last century. Although Dewey's philosophy is, like Husserl's, experience-centered, it was clearly not consciousness-centered. I begin with his observations in "Psychology and Philosophic Method" (1910): "If the individual of whom psychology treats be, after all, a social individual, any absolute setting off and apart of a sphere of consciousness as, even for scientific purposes, self-sufficient, *is condemned in advance*" (my emphasis).[10] Placed in juxtaposition to Husserl, this take on consciousness fundamentally denies the successful possibility of a phenomenological reduction to a pure consciousness as, at best, an abstraction. Dewey goes on, in his critique of psychology—echoed in a different way by Husserl's critique of psychologism—to say that "psychology is a political science. . . . The statement . . . is that psychology is an account of consciousness *qua* consciousness . . . without troubling itself with what lies outside."[11] Dewey goes on to conclude, "'Consciousness' is but a symbol, an anatomy whose life is in *natural and social operations*" (my emphasis).[12] Now, although such a statement clearly identifies Dewey as a modernist, a secondary question is less clear—does this commit him to naturalization in philosophy?

Dewey himself seems to accept this commitment: "It has appeared to me that the notion of experience implied in the questions most actively discussed gives a natural point of departure."[13] I argue that such a naturalization lies in Dewey's adaptation of a Darwinian organism/environment model of experience taken in much the same rhetorical way that Husserl's taking of the transcendental tradition of subjectivity was taken. And what is interesting is the outcome in the interpretation of organism/environment:

Dewey, in his own way, inverts the early modern epistemology of a private subjectivity. His favored victim, however, was more often Locke than Descartes. He notes, "In the orthodox view, [empiricist] experience is regarded primarily as a knowledge-affair."[14] But Dewey, echoing and transforming his favored organism/environment model, shifts the notion of experience toward action or practice: "But to eyes not looking through ancient spectacles, it assuredly appears as an affair of the intercourse of a living being with its physical and social environment."[15]

Now, an organism/environment actional model, Dewey characterizes as *experimental*. "Experience in its vital form is experimental, an effort to change the given; it is characterized by projection, by reaching forward into the unknown; connection with a future is its salient trait."[16]

Whereas Husserl, too, argued for an inner time structure in intentionality, the Deweyan twist drives his notion of temporality in a Heideggerian direction—toward the future. "We live forward; since we live in a world where changes are going on whose issue means our weal or woe; since every act of ours modifies these changes and hence is fraught with promise, or charged with hostile energies . . . adjustment of the organism to environment takes time in the pregnant sense; every step in the process is conditioned by reference to further changes which it effects. . . . Anticipation is therefore more primary than recollection; projection than summoning of the past; prospective than the retrospective."[17] Then, finally, Dewey in his version of the interactive implied ontology here suggests a different variant of what Husserlians could possible recognize as passive and active syntheses: "Experience is primarily a process of undergoing; . . . of suffering and passion, of affection. . . . The organism has to endure, to undergo, the consequences of its own actions. Experience is no slipping along in a path fixed by inner consciousness. Private consciousness is an incidental outcome of experience of a vital objective sort; it is not its source. Undergoing, however, is never a mere passivity. The most patient patient is more than a receptor. He is also an agent—a reactor, one trying experiments, one concerned with undergoing in a way which may influence what is still to happen. . . . Our passivity is an active attitude. . . . Experience, in other words, is a matter of *simultaneous* doings and sufferings."[18]

Before discussing what I shall hold is a later convergence between Dewey and Husserl in a complementary elevation of an interrelational ontology, I want to point to a different sense of "world" in Dewey, which I do not think can be found in Husserl. If Dewey remains in however limited a sense a naturalizer, his naturalism revolves around this sense of a more resistant world than the one found in Husserl. "One of the curiosities of orthodox empiricism is that its outstanding speculative problem is the existence of an 'external world.' For in accordance with the notion that experience is attached to a private subject as its exclusive possession, a world like the one in which we appear to live must be 'external' to experience instead of being its subject matter. I call it a curiosity, for if anything seems adequately grounded empirically it is the existence of a world which resists the characteristic functions of the subject of experience; which goes its way, in some respects, independently of these functions, and which frustrates out hopes and intentions."[19]

In tone, I wonder if Husserl's world remains closer to that of early modernity than Dewey's resistant world. By using consciousness as my variable, I have now shown a tension between the two most important philosophies of experience in the twentieth century: phenomenology and pragmatism. Which is more radical? Which is more true to experience? Both, I argue, have similar aims in attacking and deconstructing early modern epistemology, whether Cartesian or Lockean. And both end in substituting an interrelational ontology for the subject/object, inner/outer, dimensions of that earlier epistemology. Yet Husserlian phenomenology seems still caught in a philosophy of consciousness and Deweyan pragmatism in at least a quasi-Darwinian environmental interrelational mode.

A Lifeworld / Nonreductive Naturalization Convergence

I do not want to leave this chapter with a mere juxtaposition of these early twentieth-century philosophies of experience. So I can only suggest what I see as a convergence of phenomenology and pragmatism that I find in Husserl's late move to notions of a lifeworld, and the suggestion that the naturalization that adheres to Deweyan pragmatism is ameliorated by the contemporary direction toward a nonreductive naturalism.

I can only hint at this convergence, first from Husserl: It is in the *Crisis* that Husserl most fully develops the sense of a lifeworld, both historical-cultural and natural. And although the vocabulary of subjectivity remains, there is a marked shift to a praxis orientation. This is particularly strong in the associated "Origin of Geometry" appendix, in which geometry has its origins traced back to measuring practices, and also in Husserl's notion that writing materializes language in a way that recovery of previous space-times may occur. And finally, one may note in the *Crisis* that the elaborate step-by-step scaffolding of reductions— which are rules of interpretation that deconstruct Cartesian metaphysics—are replaced by a permanent universal epoche. "It is to be noted also that the present, the 'transcendental' epoche is meant, of course, as a habitual attitude which we resolve to take up once and for all. Thus it is by no means a temporary act, which remains incidental and isolated in its various repetitions."[20] This shift to praxis, to the development of skills leading to higher cultural achievements— such as geometry—and the permanent taking of a world as historical-cultural-natural, I see as a convergence toward its counterpart in Deweyan pragmatism. Husserl's lifeworld is both more dynamic and its inhabitants more experimental in their practices than those of the previous Cartesian setting.

Reversing the perspective, one should note the fate of early Deweyan naturalism. Whereas Husserl's philosophical fate was, within the phenomenological trajectory, to be transformed in both hermeneutic (Heidegger, Gadamer, Ricoeur) and existential (Sartre, Merleau-Ponty) directions, Dewey's fate was transformed by his later analytic followers (Davidson, Quine, and Rorty on one side, Dennett and others on another). And in my reading of Dewey's naturalization directions, taken in a modified form from Darwin, I see no strong hint of naturalist reductionism anywhere, although with some of his analytic followers it is precisely a naturalizing reductionism that obtains. And it is here that I return to the new fashionability of consciousness in both humans and animals.

Reductionist naturalization is familiar to all of you well read in the "brain-in-a-vat" school of analytic philosophers of mind. Reductions of consciousness, mind, and inner states to brain states determined strictly by electrochemical neurological processes are the strong versions of the reductionist form of naturalization. Phenomenologists should beware this direction, I agree. But that is also the reason that the animal cognitivists may hold much more promise for a phenomenological-pragmatist convergence. The lessening of the animal-human divide is also a reduction of reductive naturalization. It is a major paradigm shift for animal ethology to now recognize cognitive, quasi-linguistic, and other mental processes among our animal relatives. And it is a paradigm shift in its Kuhnian sense insofar as the new model—animals with minds, cultures, and technologies—changes how and what counts as scientific observation of these behaviors. It also entails, at a deeper and less explicit level, a challenge to the long held Euro-American prejudice against a too close animal-human parallelism. So I close by returning to my remark about corvid intelligence and crows, jays, and ravens as Deweyan pragmatists. The intelligent behaviors of smart animals, including corvids and primates, can now be described as experimental and filled with precisely the consequentialist anticipations noted in beings whose consciousness is fully temporal.

6

Adding Pragmatism
to Phenomenology

Pragmatism and phenomenology are twentieth-century American-European philosophical twins. Both began early in the twentieth century and have undergone modifications that still hold in the twenty-first century. Here I examine both Deweyan and Husserlian beginnings with a look at their later developments and with an eye to postphenomenology.

Previous scholarship has pointed up many of the interconnections between pragmatism and phenomenology: Husserl was strongly influenced by William James, whose principles of psychology he had already digested in the 1890s, and he credited James with showing him the way out of "psychologism." Bruce Wilshire's *William James and Phenomenology* (Indiana, 1968) thoroughly traces this relationship. Charles Sanders Peirce also plays a role in what are sometimes called the *structural phenomenologists* of the Göttingen School. Elmar Holenstein has traced this relationship in *Roman Jakobson's Approach to Language: Phenomenological*

Structuralism (Indiana, 1976). Later, the parallelisms between George Herbert Mead's "symbolic interactionism" and
the work of Alfred Schutz, influenced the famous "social
construction of reality" work of Berger and Luckmann,
both of whom acknowledge the debt.

That leaves John Dewey as the remaining pragmatist prin-
cipal. And it is to Dewey that I turn in my adaptation of
both pragmatist and phenomenological styles of analysis. I
do so, however, not by way of the usual scholarly comparisons but in an autobiographical way. Before doing this, I acknowledge that the scholarly approach has been partially
undertaken by others, most notably by Carl Mitcham in his
Thinking through Technology (Chicago, 1994). This book,
usually considered the most definitive history of the philosophy of technology, contains an explicit section on the work
of John Dewey and my work. Mitcham calls this a "pragmatic phenomenology of technology," and holds a considerable discussion of its main features, including Dewey-Ihde
comparisons. There are others, but I will mention only the
contribution of Peter-Paul Verbeek in *American Philosophy
of Technology* (Indiana, 2002), who also takes note of my
"pragmatism" in comparison with Heidegger.

Autobiography—from Phenomenology
to Postphenomenology

I have learned, from a now forty-plus-year history of published books, that if one accepts the title of "phenomenologist," prices have to be paid, as previously noted. These tolls
are of two main types: The first type is what I would call
a *misunderstanding toll* and arises from early criticisms of
phenomenology. For example, phenomenology is usually
thought to be a subjectivist philosophy. Another criticism is
that phenomenology is usually thought to take as evidence,

or its knowledge base, that which is "intuitively given," and thus it is associated with a form of naïve realism. And a third criticism is that because it is subjectivist and presumably relies on intuitive experience, its method is believed to be "introspective."

Although I have always denied each of these interpretations of phenomenology, I also have to recognize that some of Husserl's language lends itself to this misinterpretation, particularly since he adopted versions of both Cartesian and Kantian terminologies. If one does not recognize that even when he uses these terminologies, he actually inverts them, nonetheless the misinterpretation is bound to occur. I also have to recognize, now, after so many years, that these misperceptions will not simply disappear, regardless of how ill-informed they may be. I have addressed this in some degree in a the chapter titled "If Phenomenology Is an Albatross, Is Postphenomenology Possible?" in *Chasing Technoscience: Matrix for Materiality* (Indiana, 2003). And I shall take this problem up again below.

The second toll charged to phenomenology, is what I shall call a *freeze-frame* notion of phenomenology. It is the notion that phenomenology—whatever it is—remains static and usually is simply associated with Husserl. I think this is the version of phenomenology held by Richard Rorty, and made popular within the Anglo-American community of philosophers! Both in his masterly *Philosophy and the Mirror of Nature* (Princeton, 1979) and *Consequences of Pragmatism* (Minnesota, 1982), Rorty repeatedly characterizes phenomenology as "foundational," belonging to the outmoded metaphysical systems of philosophy, and associated with the styles of both Descartes and Kant, which he rejects. Rorty divides philosophy into groupings, and those who hold metaphysical systems, essentialist positions, in modern times take philosophy as a science, and the like are all rejected—called "idealist."

Thus Rorty says in *Consequences of Pragmatism,* "I myself would join Reichenbach in dismissing classical Husserlian phenomenology, Bergson, Whitehead, the Dewey of *Experience and Nature.* . . . [All of these] seem to me as merely weakened versions of idealism."[1] I am unaware that Rorty has ever considered any version of phenomenology as other than a freeze-framed, foundational philosophy.

As early as 1984 I began to contest this interpretation of phenomenology in a set of lectures in Göteborg, Sweden, later published as *On Non-Foundational Phenomenology* (Fenomenografiska notiser 3, 1986). That same year my *Consequences of Phenomenology* (SUNY, 1986) appeared as well, an obvious counterpart to Rorty's earlier book. My later *Postphenomenology: Essays in the Postmodern Context* (Northwestern, 1993) again took up these issues, including responses to Rorty.

What I first called a nonfoundational phenomenology, and later retitled postphenomenology, is in effect, precisely a pragmatic phenomenology. To show how and why this is the case, I shall return to the pragmatist traditions, and more precisely the Deweyan version of it, to show how this works.

A Deweyan Frame

I shall partly follow Rorty here again. In his *Consequences of Pragmatism,* he says, "My first characterization of pragmatism is that it is simply anti-essentialism applied to notions like 'truth,' 'knowledge,' 'language,' 'morality,' and similar objects of philosophical theorizing."[2] Anti-essentialist, nonfoundational, implicitly antimetaphysical—all describe pragmatic characteristics, according to Rorty. This negative demarcation can also be applied as an opposition to the specific set of philosophical traditions mentioned, making pragmatism anti-Platonic, anti-Cartesian, and anti-Kantian; indeed, all these philosophers are frequently mentioned by

Rorty as well. *But, I contend, it is also this tradition that is rejected by phenomenology—and even by Husserl himself.* This rejection, however, is sometimes not clearly apparent unless one recognizes that Husserl used both Descartes and Kant only to *invert* them. I do grant that an inversion can "mirror" and thus fail to make clear the escape sought.

However, when Rorty's negative characterization is made positive, we can begin to see something different. "Rather, the pragmatists tell us, it is the vocabulary of practice rather than theory, of action rather than contemplation, in which one can say something useful about truth."[3] And although this shift to praxis may have occurred in early pragmatism, today it is a common mode of analysis in phenomenology and all the way into contemporary science studies and philosophy of technology. To be sure, the explicit use of action, practice, and historical sedimentation comes late to Husserl—in the *Crisis*—but it is there from the beginning in the existentialization or embodiment phenomenology of Merleau-Ponty and equally in the moves made by the early Heidegger.

Rorty puts what is implicitly the whole foundationalist tradition this way: "So pragmatists see the Platonic tradition as having outlived its usefulness. This does not mean that they have a new, non-Platonic set of answers to Platonic questions to offer, but rather they do not think we should ask those questions any more."[4] From my perspective each of these notions may be found in phenomenology as well, and if we can debate about how much this is so with Rorty's freeze-framed version of phenomenology, I would claim it is clearly so in postphenomenology.

Let us look briefly at a few implications that follow from the rejection of essentialism/foundationalism and the emergence of a praxis orientation. Because of my interests in technoscience, I shall look first at what would have been classically termed *epistemological* concerns. Negatively, what are usually termed "correspondence theories of truth" entail a picture of truth that follows a trajectory from Plato's copy theories, through Descartes's representationalist theory, right up into what remains still alive in today's analytic cognitive science theory—brain imaging that purports to locate "where" the representations are located in the brain. Pragmatism—and including its analytic version in Rorty's preferred list of Quine, Sellars, Davidson—replaces representationalism with varieties of nonrepresentationalist epistemologies. But so does phenomenology. And whether or not nonrepresentationalism can be found in Husserl, it is surely found in both Heidegger and Merleau-Ponty, and continues in the more recent work of Hubert Dreyfus as well as in my work.

In *Consequences of Pragmatism,* Rorty lists two other characteristics of pragmatism, the denial of the "is/ought" and "fact/value" distinction, which while also appearing in phenomenology, is not of my interest here, and last Rorty claims that the doctrine that there are no constraints offered by foundational accounts of the nature of objects, minds, or even language in pragmatism, leads, in effect to a *coherence* version of truth results, "there are no constraints on inquiry save conversational ones. . . . . [There are only] constraints provided by the remarks of our fellow inquirers."[5] Here we arrive at Rorty's famous "conversations." On this point I simply disagree with Rorty and will show why below.

Pragmatism Up to Date

To this point I have been looking mostly at origins and characteristics; now it is time to look at a more contemporary scene. Classical pragmatism has obviously undergone considerable transformation and change. Rorty's *Philosophy and the Mirror of Nature* was a well-discussed demonstration of that with regard to analytic, now sometimes called postanalytic philosophy. Rorty's favorites, Quine, Sellars, David-son, all become "pragmatic" analysts precisely by dissolving all the essentialist, foundationalist, and older traditions already mentioned. They do so by rejecting the grounds for any representationalist and correspondence notion of truth, by dissolving transcendental/empirical distinctions, and by developing nonfoundational understandings of lan-guage, "not as a *tertium quid* between Subject and Object, nor as a medium in which we try to form pictures of reality, but as part of the behavior of human beings."[6]

But postanalytic pragmatism retains its linguistic turn and remains primarily a language-focused philosophy—this applies to Rorty as well with his adaptation of "conversa-tions," and if we look back at the emphases already noted—even with respect to practices—the postanalytic focus is on "the *vocabulary* of practice . . . of action . . . etc."[7]

If I grant Rorty's interpretation of how analytic philoso-phy, through its nonfoundationalists, Quine et al., reclaimed the pragmatist style—and I think Rorty was right in this—then my version of postphenomenology does precisely the same thing with and through phenomenology. I, too, reclaim the pragmatist style, but in a distinctly phenomenological way. This will call for differentiating a postphenomenological pragmatism from a postanalytic one. Such a nonfounda-tionalism is not tied to the "linguistic turn"; rather it is tied to a perceptual-embodiment praxis turn. Dewey's nonre-ductionist naturalism helps. For this I return once more to Dewey and his way of escaping the Cartesian trap.

In one sense, one can say that Descartes (as one variant of the foundationalist tradition being discussed) "invented" the "subject," the "object," and its division into "inner" and "external." Private, internal experience is that which is sub-jective, in the camera obscura model Descartes used, and this is separated from external and objective things outside the camera box.[8] Dewey recognized this move as a reductionist one and posed the following response:

> When objects are isolated from the experience through which they are reached and in which they function, experience itself becomes reduced to the mere process of experiencing, and experiencing is therefore treated as if it were also complete in itself. . . . Since the seventeenth century this conception of experience as the equivalent of subjective private experience set over against nature, which consists wholly of physical objects, has wrought havoc in philosophy.[9]

In short, Dewey, restored a robust notion of "naturalized" experience in which its context and relationships are reestablished, both undercutting and escaping early modern epistemology. But—this is also exactly what Husserl did with the notion of intentionality and his multidimensioned, actional notion of experience. The implications of this robust notion of experience, however, only began to be clear in the later Husserl and more prominently in Merleau-Ponty. And those implications are what I shall call *whole body experience* with the full panoply of actional perception. (I called this the praxis-perception model in *Instrumental Realism* [Indiana, 1991].) To put it simply, if postanalytic pragmatism retained the focus on language, the linguistic turn, then postphenomenological pragmatism focuses on the role of embodiment in relation to a lifeworld or the experienced environment.

Phenomenology Enriching Pragmatism

To this point I have taken account of connections between phenomenology and pragmatism, primarily with a view to the ways in which pragmatic nonfoundationalism, nonrepresentationalism, and yet retention of nonsubjective experience may be adapted into a postphenomenology. It is now time to turn the direction of the analysis to show how phenomenological techniques—or what Dewey might have called tools—can enrich pragmatism. I shall here primarily follow my own developments of phenomenology.

First, experience. Although I retain the notion of intentionality from classical phenomenology, I take it to be much more like the Deweyan concept of a contextual, interrelational process akin to an ecological organism-environment notion, or, alternatively, a relativistic situation in physics whereby the relativity between positions must be taken into account (see *Technology and the Lifeworld* [Indiana, 1990]). I reject entirely the Cartesian "homunculus-in-a-box" notion whereby all that I have are my sensations, subjectively. In short, my notion of experience is nonsubjectivist.

Second, variational theory. I place the phenomenological practice of using variations as my primary tool for analysis. Such a variational practice works well to establish both the richness of variety provided in lifeworld experience and to locate whatever structural features may be found. The best known of my analysis through this practice is *Experimental Phenomenology* (SUNY, reprinted 1986, and as a second edition, SUNY, 2012), where I show a variational range to multistable visual phenomena. A more recent example of this is my phenomenological variation on the popular Wittgenstein and Kuhn "duck-rabbit." I simply show that instead of a simple bistability, the multistable figure can be taken as (a) a duck, (b) a rabbit, (c) a squid, or (d) a Martian—all

Duck   Rabbit   Squid   Martian

Figure 2. Multistable doodle

equal variations but perhaps previously unrecognized (see Figure 2). Phenomenological variational expansion usually produces a richer and more robust notion of most phenomena.

Multistability. What variational practice often shows is a nonreductive and multistable pattern to various phenomena. In the more concrete areas of human anthropological and cultural variations, for example, one can show the multiple solutions humans have worked out with respect to a wide variety of problems. Again, my best-known example is one that relates to the variant navigational practices worked out with very different systems between European and South Pacific sailors. See *Technology and the Lifeworld.*

Embodiment. As indicated above, as phenomenology matured, from the late Husserl on, into Merleau-Ponty in one direction, and through Heidegger in another, the emphasis on embodiment in the first, and a hermeneutics of history and culture in the latter, even further removed phenomenology from the metaphysical traditions noted above. I have also developed both these emphases, but in relation to *technology.*

Critical hermeneutics. The interpretation style of phenomenology follows from the hermeneutic traditions. Heidegger, but also Paul Ricoeur, Hans-Georg

Gadamer, and others have frequently welded together hermeneutics and phenomenology. An illustration of my own will follow below.

Such are the major outlines and configurations of what I have called a *postphenomenology*. And before illustrating how such a pragmatic phenomenology works, I want to circle back to the tolls I mentioned earlier that anyone remotely identified with "phenomenology" has to pay. Regarding "subjectivism," as early as Husserl's intentionality, particularly as modified existentially as "being-in-world", the subject-object split and relation was presumably abandoned. Rather, something more like organism-environment with both "private" and "public" features operates instead. Then with respect to the "intuitively given," what variational practice accomplishes is precisely the reverse of this "myth of the given." Rather, variational practice shows *how something can become "intuitive."* The duck-rabbit-squid-Martian variation shown above produces new intuitions.

All these examples from phenomenological practices demonstrate tools that early pragmatism did not have, but they are tools that can enrich results in contemporary phenomenological pragmatism. The two strands are complementarily compatible. To this point I have now outlined a set of interrelations between pragmatism and phenomenology as closely related contemporary philosophical styles. Here the context is philosophy of technology, so the next step is to show how a pragmatic phenomenology, or postphenomenology, can deal with technologies.

Questions of Technologies

Two preliminary acknowledgments should be made: First, with respect to classical pragmatism, it is clear that Larry

Hickman was the person who reintroduced John Dewey to philosophy of technology with his groundbreaking book, *John Dewey's Pragmatic Technology* (Indiana, 1990). And following that tradition himself, he also published *Philosophical Tools for Technological Culture* (Indiana, 2001), in which he develops further a pragmatist philosophy of technology.

Second, my own contributions to philosophy of technology, now going back more than forty years, were a direct development from the phenomenology I refer to above. In contrast to Husserl, who rarely referred to technologies except for his insights about writing and measurement practices, Merleau-Ponty and Heidegger suggested a possible line of development. I have referred to this work that includes an analysis of Merleau-Ponty's blind man's cane and the feather-hatted woman, and extensively with respect to Heidegger's tool analysis (see *Technology and the Lifeworld*). Following this lead, I early developed my "phenomenology of technics," which describes a series of human-technology relations and which is widely anthologized. In short, from the seventies on I included technologies in the consideration of human experience in its pragmatic-phenomenological sense. I shall not review these here.

Instead, I want to do one short pragmatic-phenomenological analysis of new technological developments to demonstrate how such an approach works in philosophy of technology. Note, in the heading of this section, rather than *The* question of technology, which would have been Heideggerian, I more modestly and more pragmatically deal with "questions of technologies."

The technologies I wish to question are the radical, new imaging technologies that began to be developed in the mid-twentieth century and which today are radically transforming the sciences that use them. I find these technologies highly philosophically provocative because imaging raises all the right epistemological questions about representational versus nonrepresentational knowledge. These imaging technologies raise questions about constraints posed by human experience and embodiment; and they raise questions about the role of a *hermeneutic*-style philosophy that arises in both pragmatist and phenomenological traditions. Preliminary work can be found in both *Expanding Hermeneutics: Visualism in Science* (Northwestern, 1998) and *Bodies in Technology* (Minnesota, 2002).

Hermeneutics in Technoscience: From Black Holes to the Greenhouse Effect

I begin with a brief look at new instruments in astronomy and the earth sciences. First, astronomy: The production of astronomical knowledge is, of course, ancient. Virtually every ancient people developed extensive knowledge of such phenomena as sun and moon cycles, the seasons, solstices, and many other still valid knowledges. Moreover, they did this not simply with naked observation, but with measuring perceptions mediated technologically. Stonehenge in the UK, sunline tunnels in Mexico, Babylonian cuneiform tablets in Iran—all describe and record this ancient knowledge. Especially impressive is the ancient work on astronomy in China. Star catalogues were developed nearly 3,000 BP; the exploding supernova, Crab Nebula, of 1054 was noted by the Chinese (but also noted nearly worldwide, including in North America by Amerindians); and although the technology of the telescope did not take shape in China, other astronomical technologies did: sighting scopes, armillary spheres, and accurate star calendars.

But modernity in astronomy begins with the development of the telescope. Early modernity raises the stakes with the invention of lenses, another mediating technology, which, through gradual improvement, in fact dominates early modern astronomical science until the twentieth century. But, from the seventeenth century to the twentieth, astronomy was primarily conducted with optical instruments limited to the wave bands of light.

First pragmatic-phenomenological response: In the full environmental or lifeworld contexts there were clearly a wide variety of different human concerns into which this knowledge production fell or was related: practical concerns such as planting cycles and fertility issues, religious concerns such as cyclical patterns for ordering society, even political concerns such as wars and recrowning kings. But this does not mean that the knowledges produced of the cycles and such are invalid; although anachronistically brought into the present, they do get recontextualized in a contemporary scientific culture. Then from the variable of technologies and embodiment (praxis and perception as noted above), the humans were experientially relating, through sedimented and disciplined practices, what they could perceive within a technological instrumental means. Indeed, one further product was the proto-writing that inscribed cycles on reindeer bones at least as far back as the last Ice Age.

With early modernity and lenses, a new phenomenon was introduced by way of technologically mediated vision. I have previously and extensively phenomenologically described both telescopic and microscopic practices and relations; suffice it to say here that the technological transformations that lenses produced—magnification, the production of apparent distance with its new body-distance experience, and the possibility of newly visible phenomena such as Galileo's moon mountains and the satellites of Jupiter—led to an explosion of knowledge production but also situated it within the complex historical and social contexts of the time. Yet however radical early modernity was with its new instruments, it remained within a comparative, recognizable analogue practice that included human vision. Galileo's telescopic moon remained easily identifiable with the nontelescopic human sighting, although one could see some reason, given the new features revealed, why Galileo ended up arguing that telescopic vision was better than naked-eye vision.

The third and possibly "postmodern" astronomy begins, I am suggesting, in the twentieth century. In briefest form: Radar and radio astronomy were developed from World War II technologies and eventually led to the discovery of many "radio sources" that did not always overlap with optical emissions. In addition the background radiation of the universe itself was discovered (a Nobel for this was awarded, finally, in 1978 to Robert Woodrow Wilson). There has been a dramatic shift from all earlier astronomy to radio astronomy and today's astronomy. Contemporary astronomy gets its "vision" mediated by a range of instruments sensitive to gamma to radio wave emissions, with the result that merely modern optical astronomy appears to be highly limited.

Here the pragmatic-phenomenological response must be one that takes into account this new situation. The evidence for gamma to radio emissions is typically an image. The spectrum itself—were it imagined to be like a piano keyboard with eighty-eight keys—would image wave radiation or frequencies such that optical light would be represented by just two keys on each side of middle C. All the remaining keys represent the wave frequencies from sound to gamma waves. The instruments that detect this radiation may take the shape of a data projection, or, equivalently, a depiction in an image. The various technologies—instruments—that produce these results are many, complex, and variable. Many produce what I shall call frequency slices. For example, one can image the X-ray frequencies of the spectrum as the

Chandra Source does; or, one can image radio sources; or, one can image either infrared or ultraviolet (close to the optical spectrum).

Each of these new technologies mediates the human experience from its positioned and perceptual embodiment as lenses did before—but now with a difference, since what is being imaged cannot be directly experienced and thus is not like Galileo's moon. Rather, a second process must come into play, a process I call a translation. The radio "telescope" translates its long frequencies into a visualization that, in turn, can be "read" or hermeneutically interpreted. This interpretation, since it is a visual pattern, can bring into play all the advantages of human visual perception (gestalt recognition, all-at-onceness, etc.). But—and here the going gets complex—this image is always less than and more than a "picture." To be understood, it must be critically, hermeneutically interpreted. (Similarly and in parallel with this claim, note that the same thing occurs in medical imaging: One must learn to "read" X-rays, MRIs, PET scans and the like.) In short, contemporary science in its complex technological immersion also calls for active bodily, hermeneutic actions or practices.

In my astronomical examples, the new beyond bodily capacity sensing can undergo a whole series of variations (I have long called these *instrumental phenomenological variations*), which can vary between slices and composites. Slices and composites reveal quite different things, for example, in the X-ray slice of the Crab Nebula that shows its central pulsar and the radiation jets spitting out each axial end, or its overall shape as in the composite (see Figure 3).

The unseen and so far unmentioned technology that makes all this possible is, of course, the computer or, better, computer processes. Computer processes make possible the transformation of data into images or of images into data,

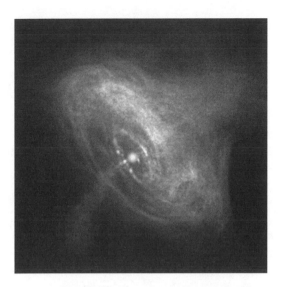

Figure 3. Crab Nebula. Photographed from Hubble Space Telescope.
Image courtesy of NASA/ESA/ASU/J. Hester.

one of the postmodern capacities of contemporary technologies. But this interchangeability also can produce something that is even beyond the capacities described to this point. Computer processes can also simulate. The first useful simulations go back, again, to World War II when simulations were used in the development of the atomic bomb in the Manhattan Project. I shall not rehearse this already well known history, but instead, go to the sciences that are most highly dependent on simulation technologies: the earth sciences, particularly those related to questions of global warming (see Chapter 3).

All of the evidence, although not all produced by simulations, gets plugged into simulation processes that produce the evidence for global warming. If one goes back to my previous example, images "look" like pictures, that is the analogues are map-like, whole earth displays with the oceans colored with different shades. To hermeneutically "read" these depictions, you need to know that the colors are "false colors" or colors assigned with designated associations with some kind of intensity scale. One might want to think that these show temperature variations—but as you can see from the script, that is not the case. It is rather a modeling—from "real data"—of ocean levels, and the whole process purports to show that the gross level of the oceans rose by 1.6 mm between 1993 and 1994.

I have sometimes claimed that if these are "pictures," then they are the most expensive pictures ever produced. The production of the data on which the image is based includes satellite measurements from multiple passes, land measurements, oceanic floating devices, and the like, multimillion-dollar instrumentation. All this is then fed into the computer process and simulation, which uses tomographic procedures to produce the single depiction.

This is a glimpse at the science involved, a complex, corporate, Big Science, international in scope and geographically encompassing. And the science meets global politics. Does, or even can, the simulation accurately portray global processes? (shades of correspondence theory). Or if it does, how much is homogenic? And even if so, how much can we affect? All this brings us to the current controversy.

Too bad, in some respect, that this looming conclusion to my essay has brought us precisely up to where we are. But I can make a few final pragmatic-phenomenological observations at the end. First, I hope to have shown that this new version of imaging cannot fit well into traditional forms of "representation," for even so far as it is a representation at all, it is one that calls for its analysis in a critical, hermeneutic mode that itself is not representational. Second, however complex the scene, any interpretation clearly implies an embodied interpreter. The image, the scene, must have been displayed through the translation process that implies precisely our perceptual capacities and embodied possibilities. Third, none of this can be naively undertaken but must undergo the critical hermeneutic processes suggested above. And fourth, this recasting of science also replaces it within the broader and pragmatically recognizable role within human cultural and historical behavior. Both a pragmatic and phenomenological approach deny the now outdated nineteenth-century version of an ahistorical, noncontextual, value-free science. Both situate technoscience within the complex human activities noted.

So, interestingly, at the end, once again pragmatism and phenomenology converge in almost precisely the way Rorty earlier suggested: Philosophy in its "post" guise, pragmatic or phenomenological, moves from the early modern epistemology toward a critical hermeneutics. Hopefully what results will also be "edifying."

7

From Phenomenology
to Postphenomenology

The act of naming frequently follows somewhat behind the thing or process to be named. That was clearly the case with both science and technology. If early modern science began, as the standard view has it, in the seventeenth century, it was not until 1833 that it was named *science* by William Whewell. Earlier the most frequent name was *natural philosophy*. Similarly, technology for a long time, even after the Industrial Revolution, was called a product of the industrial arts, or simply *machines*, or products of engineering. Historians of technology, including Thomas Hughes and David Nye, point out that *technology* did not become common parlance until early in the twentieth century. And finally "technoscience" as a term for the close interrelation of science and technology was not coined until the 1970s by Gilbert Hottois, a Belgian philosopher of science and technology, although there were anticipations in the '60s in the work of Gaston Bachelard, a chemist and phenomenologist.

This same lagging history also belongs to postphenomenology. *Phenomenology* as a name for a style of philosophy has an older history—clearly Hegel's *phenomenology* is an early use, but in the late modern sense, that term is more often associated with Husserl. Many others in the early to mid-twentieth century, including Heidegger, Sartre, Merleau-Ponty, and Ricoeur, claimed it. And it came in various flavors or varieties: *transcendental* with Husserl, *existential* with Sartre and Merleau-Ponty, *hermeneutic* with Heidegger, Gadamer, and Ricoeur. Even *postphenomenology* had several origins. Some writers used this term to designate any phenomenology after Husserl; but more specific groups such as a small group of Heideggerian cognitivists used it; and in Australia there are those who associate postphenomenology with a nonsubjectivist and naturalized phenomenology in the work of Cornelius Castoriadis and branches of Frankfurt critical theory. These have been discussed by Suzi Adams of Flinders University.[1] But here I concentrate of the movement that came out of our Technoscience Research Group at Stony Brook University and later in several locations in Europe, particularly Denmark and the Netherlands; today it finds representatives worldwide.

First, a look at my own publishing history regarding this term: As the preceding chapter shows, Richard Rorty was—for the dominant analytic majority—the primary figure to relate phenomenology and hermeneutics to trends in what sometimes is called "postanalytic" philosophy. He recognized the importance of Heidegger and crafted a notion he called "edifying hermeneutics" (in *Consequences of Pragmatism*). Midcentury there was something of a popular game in naming the "three top philosophers" of the twentieth century. There were a significant number of variations as to who was included, but almost all these lists included Heidegger, Wittgenstein, and one-or-other of the pragmatists—either

James or Peirce or Dewey. For Rorty it was Heidegger, Wittgenstein, and Dewey. I responded positively to Rorty's work and found strong agreement with his pragmatic version of antiessentialism, antifoundationalism, and the use of an experimental style of philosophy—all of which I thought I also practiced, although in a phenomenological style (including a critical hermeneutics). As early as 1984 I gave a set of invited lectures at Göteborg University, which were published as a monograph, *On Non-Foundational Phenomenology* (Göteborg, 1986). These were responses to Rorty, but showing how nonfoundationalism would work in phenomenology. But finding this term bulky, I later shifted to the term *postphenomenology*. I published my first book with the new "post" designation, called *Postphenomenology: Essays in the Postmodern Context* (Northwestern, 1993). So now the name was established, and other publications began to use *postphenomenology*. Today there is a quite large literature using this name.

Although it is standard academic practice—and for that matter, also scientific practice—to link precedents or discoveries to earliest dated publications, other practices are also relevant. In the case of postphenomenology, the establishment of the Technoscience Research Group with its research seminar was crucial. I had been dean of humanities and arts from 1985 to 1990, and for 1990–91 had a combined sabbatical and research semester leave, which, with my family, we took in Australia with a base at the University of Sydney. On return to Stony Brook, I had hoped to set up a research group to host what was beginning to be a trickle of visiting scholars to focus on science-technology studies. STS programs are usually interdisciplinary, and Stony Brook had none. We began informally in 1994. Sung Dong Kim from Hoseo University, Korea, and Monique Riphagen from Wageningen University in the Netherlands were the first visiting scholars, and together with a small group of philosophy PhD students we began a rigorous reading program in STS- related publications. Kim came to translate *Technics and Praxis* into Korean, and Riphagen did an individual research project. By 1997 I was fortunate enough to be promoted to a Distinguished Professorship, and so by 1998 the department agreed to establish the research seminar as a permanent part of its interdisciplinary Interface program. It was a mini-STS program. By then we had a quite regular flow of visiting scholars, mostly from the Scandinavian countries and northern Europe. We opened a new MA track in technoscience studies and in short order graduated three Danes and an Argentinean. In addition, a number of PhDs, both from Stony Brook and visiting scholars, did dissertations with postphenomenological analyses. In the seminar we would read thematically, a different focus each semester, and our first concentration was on *materiality*. We decided early to read works by Donna Haraway, Bruno Latour, Andrew Pickering, and me—and this theme led to our first book, *Chasing Technoscience* (Indiana, 2003), edited by Evan Selinger and me. Later the traditions of reading only living authors, of picking a significant figure and reading deeply, and then inviting the author to be "roasted" took off. We roasted fifteen principals from philosophy of technology, science, history and sociologies of science, and feminism. Finally, part of the seminar program encouraged producing research papers aimed for conference presentations and possible publication. This led to the series of "postphenomenology research panels," which, from 2007 on, began to appear on programs of 4/S (Society for the Social Studies of Science), SPT (Society for Philosophy and Technology), and SPHS (Society for Phenomenology and Human Studies). Most recently we are collaborating with the Philosophy of Science Association with a panel of philosophers, all of whom have been on postphenomenology panels. The first such panel took place at the PSA (Philosophy of Science Association) meeting for 2014.

Results ended up as publications in all the major STS journals and now will begin to appear in the new Lexington Books series "Postphenomenology and Philosophy of Technology." This, however, was a beginning, although one that saw the technoscience seminar as a major incubator for postphenomenological research. But if Stony Brook was a beginning, postphenomenology now finds more and more centers of interest.

Parallel to our North American activities, Peter-Paul Verbeek of the University of Twente also began to claim postphenomenology, with his dissertation and books coming out of Twente, then his later *What Things Do* (Penn State, 2005) and *Moralizing Technology* (Chicago, 2011). He rapidly became successful, and since roughly 2000 has had a large group of PhDs, and he directs an MA program. He has a large northern European network that also draws researchers to postphenomenology. In Denmark Finn Olesen and Cathrine Hasse have begun another network. All of us have collaborated in forming programs that take place primarily in Europe and North America but also in Asia and South America.

At this point I turn from the anecdotal and personal histories back to some of the philosophical results in the movement from classical phenomenology to the modifications of classical into postphenomenology.

Following Husserl

As noted in the preface, I had deliberately undertaken to follow Husserl in doing phenomenology. Thus in the earliest days of doing philosophy in the phenomenological style, it was the sense of embodiment in use which, I now much later confess, was not the surprise it turned out to be. Technologies were found—and under rigorous phenomenological descriptive analysis—to be mediational means of experiencing an environment or world. They were not primarily experienced—except in non-use contexts—to be objects or things. The surprise about this came so much later, when introduced to the *Nachlass* and the discovery that Husserl continued to regard instruments as instrument-things (with some praxis exceptions as noted)!

As mentioned in the preceding chapters, I was far from the first to make the embodiment discovery—both Heidegger and Merleau-Ponty described material "technologies" in use as experientially embodied. Thus can one say that those who followed Husserl's advice discovered something different from his own result?

The discovery of *multistability* is the second major follow-Husserl-to-a-different-result surprise. Husserl's investigative method, patterned on mathematical variational analysis, was the use of what he called "imaginative variations," for which the result was supposed to be to determine *invariants* or *essences*. As argued, variational theory, in my estimation, is what gives phenomenology its rigor. But, again following Husserl, this time first in the first edition of *Experimental Phenomenology*, what I found was not a stable essence as Husserl also called his result, but multistability. The first cases in *Experimental Phenomenology* were most frequently called "ambiguous drawings," that is, two-dimensional line drawings that took on multiple appearances. Early phenomenology and especially gestalt psychology were familiar with this class of visual phenomena. Using notions such as figure/ground and reversals thereof, two- and three-dimensional projections, and such: "Duck/Rabbits," "Face/Vase," and "Young/Old Ladies" were simple one, two, at most three multistable appearances to a single drawing. *Experimental Phenomenology* simply took a much more radical application of variational theory and demonstrated that most two- and even three-stability figures could go to more. The famous Necker Cube figures—usually at most three different perspectives in gestalt and classical psychologies—were taken to five stabilities in *Experimental Phenomenology,* and circles with squiggle lines were taken to nine or ten stabilities. The earlier multistable doodle, which goes back to a Wittgenstein conference in the '70s at Kings College, London, has become something of a calling card (see Figure 2). This work was done in the early '70s before the sciences began to develop their own multistable analyses. By the '80s, however, multistability began to appear in many sciences.

Close to home, John Marburger, a physicist, Stony Brook president, and my colleague in an NSF-funded teaching program, had discovered optical multistabilities in the mid-1970s, precisely at the same time I published my first work on this phenomenon—but I was unaware of his work then. By the '80s thin film microchemistry discovered that multiple stacking arrangements, or multistabilities, were possible. Tetracyanoethyelene takes five shapes, and some forms of carbon take thirteen. The millions of galaxies known today (there was only one until 1924 and Hubble's discovery that Andromeda was a galaxy) have large but finite numbers of shape-stabilities (spiral, lenticular, elliptical, starburst, butterfly, dumbbell . . .), and even infinities have multiple mathematical shapes (flat, humpback, bubbly, hyperboloid . . . .). And similarly, in biology and nanomaterials this phenomenon is now well known. A recent google search for multistability in science showed 12,300 entries.

And when it comes to technologies, I would argue that multistability is virtually the norm. I have published two postphenomenological studies in the second edition of

Experimental Phenomenology (2012), one showing how seven to eight variations on the camera obscura, in the history of science instruments, account for a vast number of optical imaging technologies (photography, spectroscopy, interferometry, holograms). In another study, I show how the bow-under-tension has produced multiple trajectories of technologies in multiple cultures (archery, musical instruments, tools). In anthropology and sociology this phenomenon is also well known if not with this nomenclature (kinship patterns, cultural variants on marriage practices . . .). With respect to classical phenomenology and postphenomenologies, if the former seeks essences, the latter finds multistabilities. Thus postphenomenology in contrast to classical phenomenology finds instrumental transparency rather than object-character in technologies, and frequently multistabilities rather than essences to emerge from variational analysis.

As postphenomenology matured, in the group setting of the technoscience group and its research seminar the nuanced modifications of classical phenomenology began to become clear. Here I look at "intentionality" in its classical and Husserlian form and in its modified postphenomenological form. Intentionality, I hold, is a form of *interrelational ontology*. Husserl's formulation in the *Cartesian Meditations* is classical and clearly consciousness-centered. Here the formula is: *ego-cogito-cogitatum,* or an ego subject-thinks-something. Note in passing that both Heidegger and Merleau-Ponty modified intentionality in an existential direction. For Heidegger it is Dasein, being-in the-World, and for Merleau-Ponty it is an embodied subject, *etre au monde.* In its original Cartesian form the *ego cogito* occurs inside the subject and the object is outside but represented inside in a mental image. And as I have shown, this model is deliberately taken from the metaphor of a camera obscura (see appendix). And although in the *Cartesian Meditations* Husserl seeks to change this model into the directional intentionality in which phenomena are not mental events, it is still the case that *ego* remains a subject and what is focal and forefront is *consciousness*. But pragmatism and—and closer to phenomenological home—both Heidegger and Merleau-Ponty note that there is more to experience then consciousness. Rather, as interrelational, "being outside oneself," as Merleau-Ponty puts it, gives a much stronger sense of embodiment to intentionality. For example, prominent features of embodiment are skill attainments, which occur only through practice and training, but once learned are taken for granted and incorporated into action. In this learning, what might begin as something conscious, with practice ceases to be explicitly conscious. An example is the virtuoso keyboardist—typist or pianist—who no longer needs to be conscious of every aspect of the craft. To put it simply, postphenomenology substitutes an embodied action for consciousness or subjectivity. Perception for postphenomenology is bodily and actional.

If the human embodied actor is the ego or Dasein or lived body side of interrelational intentionality, the other side is cogitatum, or the phenomenological world, *monde*. In its prephenomenological Cartesian form what is experienced is the mental event (the image inside the camera). But in phenomenology the phenomenological world is experienced as concrete materiality. Husserl's call is for phenomenology to go to the *things themselves*. And technologies can, in one restricted sense, be things, or objects in an environment—only if they are sitting there, as it were, as objects not being used. Here is where they come closest to the vestigial Cartesian descriptive analysis. But such things do not present themselves that way in use. Nor was Husserl a reductionist. A Cartesian static object-thing is ultimately reducible to a shape-extension or geometrical object. Husserl realized that objects in themselves retain multidimensionality and present themselves to the multiple dimensions of the experiencer. That was the sense of his plena from the *Crisis*. A world is a lifeworld, which, since Husserl remained a modernist, would have to be taken as both natural and cultural. Thus a technology—say an automobile—even as it sits there is not only shiny but also has certain values. My Mercedes is different from my previous Toyota in design and engineering. Both are always more than physical and visual predicates.

However, the actional term in the intentionality arc is *cogito*—in Descartes the act of thinking. Ego thinks something. But swerving away from the consciousness ego language of Cartesianism, the embodied human experiences (more than mere consciousness) something. And if that experiencing is through using an instrument (technology), something quite different than experiencing a thing takes place. As both Heidegger and Merleau-Ponty recognized, experience goes through the instrument, and as I phrase it, the technology is taken into intentionality itself—it is a means of experiencing something else. A technology occupies a mediational position between the acting embodied user and its referential and directional fulfillment in the environment. And it does so in different, multiple ways (see my phenomenology of technics in *Technology and the Lifeworld*). Consider further the example of the keyboardist, typist or pianist. Although the materiality of the keyboard remains in the selective affordances and constraints of the instrument, what is focal is the transparency-in-use that the embodied user experiences in producing a text or musical sounds. Granted, typing and piano playing call for different types and degrees of embodiment. A typist would not likely be effective were he or she to be as active bodily as a virtuoso piano player. Postphenomenology is also sensitive to the nuanced differences between such technologies—each is distinctive in the ways it may be embodied in use. For example, the typing keyboards I have used to produce books include mechanical typewriters, electric typewriters (even a battery-powered one prior to word processing), and computer keyboards—each of which calls for nuanced differences in skills. Similarly I discovered when buying a piano how different sound styles are built into various piano traditions (flowing romantic, crisp classical, degrees of resonance, etc.).

From its concrete beginnings in the technoscience context at Stony Brook and today at its European centers, postphenomenology has been dominantly an STS discipline and it has taken on several practices and patterns of this interdisciplinary field. Today these programs include a planned revival of my technoscience seminar at Stony Brook and a program directed by Cathrine Hasse in Aarhus, Denmark, titled "Future Technology, Culture and Learning," which is interdisciplinary and housed in the education department. And Peter-Paul Verbeek of the University of Twente is involved in two graduate programs, one in technology and ethics and one in philosophy and science, technology, and society. These programs are in English and those in Denmark partly in English. Both have PhD components. All three locales are heavily postphenomenological.

> For example, most STS programs include concrete case studies as part of the program. The technoscience research group has done the same, with the conviction that concrete cases—both empirical studies and postphenomenological variational practices—are more likely to descriptively catch the uniqueness of particular technoscience phenomena. Thus while retaining a rigorous use of variational theory—and,

if anything, radicalizing it—postphenomenology can push the borders of cases farther than many standard empirical studies.

Postphenomenology, more than classical phenomenology, has taken both contemporary science and technologies seriously and thus in terms of a research culture has not been criticized for being either subjectivist or antiscience. Postphenomenology retains the antireductionist stances of classical phenomenology but is open to a wider use of contemporary modeling, simulational, and statistical methods.

Again, following patterns in STS programs, postphenomenology has adopted a praxis-oriented rather than a modernist epistemological style of analysis. Thus many studies have looked at the concrete social and technical aspects of such contemporary issues in information, imaging, media, health, and ICT (information communication technologies). This pattern differs sharply from earlier philosophy of technology as well, which tended to deal with technologies *überhaupt*.

Technoscience studies are intrinsically interdisciplinary, and many postphenomenology analysts find themselves working in interdisciplinary contexts and programs. Again in contrast to earlier philosophy of science, positioning is increasingly in research and development roles rather than in the older applied roles that tended to focus upon already in-place practices and technologies. In passing I take note of recent placements of some of my own PhD advisees: Evan Selinger, Rochester Institute of Technology; Robert Rosenberger, Georgia Institute of Technology, and Kyle Whyte, Michigan State University—all are involved in interdisciplinary contexts and are re-search productive.

As noted earlier, pragmatism tends to model its interrelational ontology on the evolutionary organism-environment relation. That style of model is appropriate for postphenomenology as well, although retaining from classical phenomenology its emphasis on first person awareness in its analysis. Person-perspectives, first, second, and third, all play roles in postphenomenology.

Husserl thought phenomenology to be a new science. Its task was to examine or reexamine all the areas or regional on-tologies, and he foresaw a large group of researchers to take on this task. Postphenomenology retains that trajectory, but as noted, when the things themselves reveal themselves, it turns out they do not always conform to past notions. Phi-losophy, I argue, must itself transform itself over time. As our world, our lifeworld, changes, so must our reflections on it.

Appendix
Epistemology Engines

The increasing use of the term *technoscience* as a description of the relations between science and technology is also suggestive of other ways in which science and technology are entwined. Historians of science have a saying: "Science owes more to the steam engine than the steam engine owes to science." Historically, the steam engine developed without much explicit use of scientific theory; yet it inspired the ideas of entropy and the second law of thermodynamics. The machine, not raw nature, suggested the phenomena.

The steam engine shows how a technology can serve as a partial "epistemology engine." But a nearly forgotten optical device from early modern science gives a much fuller model for how knowledge itself is produced, a true epistemology engine. This is the camera obscura, which later evolved into the pinhole camera. The optical effect in which some external scene, under light, could be seen as an inverted image inside a darkened room on a blank screen, may have been known to Euclid, but it was clearly described by the Islamic philosopher Alhazen, in his *Optics* of 1037.

Camera obscuras, and related camera lucidas, became well known in the Renaissance: In about 1430, Leon Battista Alberti used them to trace objects with astonishing verisimilitude; the camera contributed to the development of perspective drawing. The optical effect also automatically reduced three dimensions to two. Leonardo da Vinci again described the darkroom (in about 1450) and explicitly made the camera a model for the eye: "When the images of illuminated bodies pass through a small hole into a dark room . . . you will see on the paper all those bodies in their natural shapes and colors, but they will appear upside down and smaller . . . the same happens inside the pupil."

But the camera did not become a full epistemology engine until both René Descartes and John Locke explicitly made it thus in the seventeenth century. Descartes in *La Dioptrique* and Locke in the *Essay on Human Understanding* both draw on the camera obscura as a model for how knowledge is produced. For them it is more than the eye that represents the world; the camera is to the eye as the eye is to the mind. Locke's analogues are strikingly literal: "External and internal sensation are *the* only passages I can find of knowledge to the understanding. These are . . . the windows by which light is let into this dark room: for methinks the understanding is not much unlike a closet shut from light, with only some little opening left, to let in external visible resemblances, or ideas of things without: . . . [these] resemble the understanding of a man, in reference to all objects of sight and the ideas of them."

Here we have the birth of early modern epistemology: "Reality" is "external"; knowledge is "represented" and "internal"; and "objective truth" has to be a "correspondence" between the object and its representation. But with this model of knowledge comes the problem of the inner homunculus or "subject," the self trapped inside the camera, and the need for an ideal observer who sees both what goes on inside and outside at the same time and is thus able to tell whether the object and its representation correspond. Such is the epistemology produced by the engine of the camera obscura.

This progressive history of an epistemology engine displays two movements associated with it. The first is one of escalation—from Alhazen's observation of an optical effect; to da Vinci's camera as analogue for the eye; to Locke's and Descartes's analogue of camera to eye to mind—by which the camera is made into a full epistemology engine. The second is the inward progression of the location where external reality, itself an artifact of the geometry of the imaging phenomenon, interfaces with the inner representation. For da Vinci, the interface of external/internal occurs "in the pupil"; for Descartes, it is the retina; and continuing the camera epistemology, contemporary neuroscience locates it in the brain.

As well as being an amazingly persistent epistemology engine, the ghost of this forgotten technology lurks in the "science wars." At least one dimension of this contemporary controversy revolves around a passionate defense of notions such as "external reality," "truth" usually defined as "correspondence," which is modernism's form of "objectivity," and so on. Those who have begun to question this epistemology (and its antique engine) are dubbed postmodernists and relativists, the latter of whom have yet to agree on or produce a new epistemology engine. Perhaps it is time, however, to explore a wider array of possibilities.

Acknowledgments

This reappraisal has benefited from many others. Most of the chapters were first presented—usually in shorter and less detailed form—at related conferences, particularly those of the Husserl Circle. This circle was organized in 1969, the year I moved from Southern Illinois to Stony Brook. I did not attend the first year but became a member shortly after. Questions, criticisms, and discussions were had with fellow members. As the work for this book began to be collected and revised, I frequently drew from fellow Husserl scholars, most frequently from my colleague, Donn Welton, who has written more on Husserl than anyone else I know.

Rochus Sowa, editor of various Husserliana volumes, and I became acquainted at a Husserl Circle meeting in Prague. He was extremely helpful and supplied me with the German texts concerning scientific instruments in German in 2007—the HUA volume came out in 2007. Later, Frances Bottenberg, whose first language is German and who is a Stony

Brook PhD, supplied me with English translations of the *Nachlass* materials. Søren Riis, a former visiting scholar and now at Roskilde University in Denmark, provided translations of the *Briefwechsel* and other materials. These sources draw from Husserl's posthumous publications.

Later still, after I had done a first draft of this book, I consulted several truly in-depth textual scholars, including Lester Embree and Dermot Moran. Of these Betsy Behnke provided a thorough set of references and comments from the few references to technologies and instruments in the published texts, some of which I had not known. She also did translations of some of the texts, and I have used some of these as well. I am grateful to Betsy for this help and have incorporated her information in the book. Additionally Robert Scharff supplied me with an e-file of Husserl's preface to the *Logical Investigations,* which includes observations on the microscope. Then, regarding the considerable ambiguity that surrounds Husserl's personal writing and reading habits, Rudolph Bernet and Thomas Vongehr of the Husserl Archives have provided invaluable information. All these later helpers provided information via the internet. Here one can begin to sense the network and electronic distribution technologies that are constitutive of today's technological texture.

I have already mentioned my first Husserl teacher, Erazim Kohak, then a new assistant professor at Boston University during the last of my course work related to the PhD. He was enthusiastic about Husserl, so much so that on the last day of class he removed his shirt to reveal his T-shirt, which read "The Edmund Husserl Social and Athletic Club." I was so impressed that I ordered a number of these and had my graduate students wear them while sailing with me later on Long Island Sound in my sailboat named the *Epoche*! Later we again met in Prague to round off years of occasional meetings.

Skipping to the production stage, I want to acknowledge the enthusiastic encouragement to do this book from John Caputo, the late Helen Tartar, and now Tom Lay—all from this Fordham University Press series. In between, I cannot list by name all the colleagues and students who contributed ideas to the accumulation of phenomenological variations that lie in the background of these Husserl reflections, but both the last of my Stony Brook chairs, Robert Crease and Eduardo Mendieta, helped. Earlier there were the many years of doing phenomenology in my third-floor study, and I want to acknowledge both Marjorie Miller and Elyse Glass for lessons they taught about imagination, related to experiments we undertook then. Linda helped me with some word-processing technicalities that I had not mastered. And finally again, I thank my colleagues from the Husserl Circle who discussed and criticized my Husserl papers, which are now appearing here in forms that respond to their interventions. In the years at Southern Illinois University, I met Herbert Spiegelberg, who conducted workshops from which much was learned at Washington University in nearby St. Louis.

It will be noted that I hold significant discussions about Descartes and Galileo because these two play a prominent role in Husserl's thinking. During the millennial year, 2000, both *Science* and *Nature* magazines published millennial series. I contributed to both, with two Millennium Essays in *Nature*. The appendix, "Epistemology Engines," first appeared in *Nature*, July 6, 2000. It is reprinted here with permission of the Nature Publishing Group as first published in the journal on the date indicated. Chapter 2, "Husserl's Galileo Needed a Telescope!" was first published in the journal *Philosophy and Technology* 24, no. 1 (2011): 69–82, and is reprinted here with minor revisions with the permission of Springer Publishers.

Notes

Husserliana

The following list gives the full information for the volumes of Husserliana cited in the notes. In the notes abbreviations are used, and translators for the parts cited are given. My three translators, Betsy Behnke, Frances Bottenberg, and Søren Riis, are indicated by their initials.

Husserliana 9 *Phänomenologische Psychologie.* Vorlesungen Sommersemester. 1925. Edited by Walter Biemel. The Hague. Netherlands, Martinus Nijhoff, 1968.

Husserliana 14. *Zur Phenomenologie der Intersubjectivität.* Text aus dem Nachlass. Zweiter Teil. *1929–35.* Edited by Iso Kern. The Hague. Netherlands: Martinus Nijhoff, 1973.

Husserliana 15. *Zur Phänomenologie der Intersubjectivität.* Text aus dem Nachlass. Dritter Teil, 1929–35. Edited by Iso Kern. The Hague. Netherlands: Martinus Nijhoff, 1973.

Husserliana 16. *Ding und Raum.* Vorgelesungen 1907. Edited by Ulrigh Claesges. The Hague, Netherlands: Martinus Nijhoff, 1973.

Husserliana 20/1. *Logische Untersuchungen. Erganzumsband.* Erster Teil. Edited by Ulrich Melle. The Hague, Netherlands: Kluwer Academic, 2002.

Husserliana 39. *Die Lebenswelt. Auslegungen der vorgegebenen Welt und ihrer Konstitution. Texte aus dem Nachlass* (1916–37). Edited by Rochus Sowa. New York: Springer, 2008.

Husserliana: Edmund Husserl Dokumente 3/1–10. Edmund Husserl. *Briefwechsel* [Correspondence]. Edited by Karl Schuhmann. The Hague, Netherlands: Kluwer Academic, 1994.

Introduction: Philosophy of Technology, Technoscience, and Husserl

1. Bruno Latour, *We Were Never Modern* (Cambridge, Mass.: Harvard Univerrsity Press, 1993), 13.

2. Ibid., 11.

3. Paul Forman, "The Primacy of Science in Modernity, of Technology in Postmodernity, and of Ideology in the History of Technology," *History and Technology: An International Journal* 23, no. 1–2 (2007): 1–152.

4. E-mail communication from Betsy Behnke, August 19, 2014, citing comments from Husserl's *Briefwechsel.* Behnke translation.

1. Where Are Husserl's Technologies?

1. Alfred North Whitehead, *Science and the Modern World* (New York: Free Press, 1997 paperback), 107.

2. Stephen Fuller, "Philosophy of Science since Kuhn: Readings on the Revolution That Has Yet to Come," *Choice* (December 1989): 595.

3. Larry Laudan, *Science and Relativism* (Chicago: University of Chicago Press, 1999), 183.

4. Bas van Fraassen, blurb on back cover, Ronald Giere, *Scientific Perspectivism* (Chicago: University of Chicago Press, 2006).

5. Edmund Husserl, *Ideas Pertaining to a Pure Phenomenology and to Phenomenological Philosophy: Second Book,* trans. R.

Rojcewicz and A. Schuwer (Dordrecht: Kluwer Academic, 1989), 197.

6. Ibid.

7. Ibid.

8. Harold J. Brown, "Galileo on the Telescope and the Eye," *Journal of the History of Ideas* 46, no. 4 (1985): 487–501. Brown draws from the collected works of Galileo and in subsequent notes I shall cite those sources [*opere*] as noted by Brown.

9. Galileo, *Opere,* 289.

10. Ibid., 360–62.

11. Ibid, 363.

12. Galileo, *Starry Messenger,* 46.

13. The following quotations, from Husserliana 39 are here abbreviated from the German *Nachlass* texts, translated by Francis Bottenberg. A VII 28/2a. See the list of Husserliana at the beginning of the notes section.

14. Ibid., A VII 28/3b.

15. Ibid., A VII /4b.

16. Ibid., BA 1 21/68a.

17. Ibid.

18. Ibid., B 1 14/20b.

19. Ibid., B 1 14/21.

2. Husserl's Galileo Needed a Telescope!

1. Edmund Husserl, *Ideas Pertaining to a Pure Phenomenological Philosophy: First Book,* trans. Fred Kersen (The Hague: Martinus Nijhoff, 1983), 59.

2. Edmund Husserl, *The Crisis in European Philosophy and Transcendental Phenomenology*, ed. and trans. David Carr (Evanston, Ill.: Northwestern University Press, 1970), 25.

3. Ibid., 28.

4. Ibid., 49.

5. Ibid., 354–55, 372.

6. Ibid., 48–49.

7. Ibid., 49.

8. Ibid., 382.

9. Harold I. Brown, "Galileo on the Telescope and the Eye," *Journal of the History of Ideas* 46, no. 4 (1985): 49.

10. D. J. Boorstin, *The Discoverers* (New York: Vintage, 1985).

11. Ibid., 316.

3. Embodiment and Reading-Writing Technologies

1. Friedrich Kittler, "The Mechanized Philosopher," in *Looking After Nietzsche*, ed. Laurence A. Rickels (Albany: State University of New York Press, 1990).

2. Thomas Vongehr, Husserl Archives, e-mail correspondence 12 December 2014.

3. Ibid.

4. Edmund Husserl, *Formal and Transcendental Logic,* trans. Dorion Cairns (The Hague: Martinus Nijhoff, 1977), 229.

5. Vongehr, e-mail correspondence, 12 December 2014.

6. Cited from Husserl, *Briefwechsel,* correspondence, 1934; e-mail correspondence with Betsy Behnke, her translation.

7. Husserl, *Briefwechsel,* correspondence, 1936. Behnke translation.

8. HUA 15, p. 157. BB translation. Letter, 7 October 1934, p. 105.

9. HUA 14. BB translation.

10. C 8, p. 24. BB translation.

11. Nr. 79 (C September 1931), p. 356. SR translation.

12. Betsy Behnke, e-mail correspondence, 22 December 2014.

13. Ibid., correspondence from 1933. BB translation.

14. Regarding the postphenomenological research discussed above, see *Human Studies* 31 (2008): 1–9, Special Issue on Postphenomenological Research.

4. Whole Earth Measurements Revisited

1. *Scientific American,* 24 July 2014, reported on a study that investigated seven surveys and 4,014 scientific articles concerning a consensus among scientists about global warming with homogenic causes—the consensus was 97.2 percent agreed.

From "How to Determine the Scientific Consensus on Global Warming," by Gayathri Vaidyanathan, excerpted from Climatewire, 24 July 2014.

2. Edmund Husserl, *The Crisis in European Sciences and Transcendental Phenomenology,* ed. and trans. David Carr (Evanston, Ill.: Northwestern University Press, 1970), 29–30.

3. Ibid., 35.

5. Dewey and Husserl: Consciousness Revisited

1. "The Mind of the Chimpanzee," *Science* 316 (2007): 44–45.

2. Berndt Heinrich and Thomas Bugnyar, "Just How Smart Are Ravens?" *Scientific American* 296, no. 4 (2007): 46–53.

3. Edmund Husserl, *Cartesian Meditations,* trans. Dorion Cairns (The Hague: Martinus Nijhoff, 1960), 3, 4.

4. Ibid., 12.

5. Ibid., 13.

6. Ibid., 18.

7. Ibid., 21.

8. Ibid., 30.

9. Ibid., 33.

10. John Dewey, *The Philosophy of John Dewey: The Structure of Experience,* ed. John J. McDermott (New York: G. P. Putnam's Sons, 1973), 161.

11. Ibid., 161–62.

12. Ibid., 162.

13. Ibid., 60.

14. Ibid., 61.

15. Ibid.

16. Ibid.

17. Ibid., 64.

18. Ibid., 63.

19. Ibid., 71.

20. Edmund Husserl, *The Crisis in European Sciences and Transcendental Phenomenology*, ed. and trans. David Carr (Evanston, Ill.: Northwestern University Press, 1970), 150.

6. Adding Pragmatism to Phenomenology

1. Richard Rorty, *Consequences of Pragmatism* (Minneapolis: University of Minnesota Press, 1982), 213–14.

2. Ibid.,162.

3. Ibid.

4. Ibid., xiv.

5. Ibid., 165.

6. Ibid., xvii.

7. Ibid., 162.

8. See my appendix.

9. John Dewey quoted by Rorty, *Consequences of Pragmatism,* 44.

7. From Phenomenology to Postphenomenology

1. Suzi Adams, "Towards a Post-Phenomenology of Life: Castoriadis' Critical *Naturphilosophie,*" *Cosmos and History* 4, nos. 1–2 (2008): 387–400.

Index

cell phones, 15–16
Chasing Technoscience: Matrix for Materiality (Ihde and Selinger, eds.), 105, 126
climate change, 78–79, 86–87; imaging technologies and, 120–22
coal discussion, 25–26
cognitivism, animal studies, 89–92
computerization, 63, 65
consciousness: cognitivism, animal studies, 89–92; in Dewey, 92–99; in Husserl, 92–99; intentionality, 95; language, 96; nonreductive naturalism, 100–2; psychology and, 89
consciousness-centered experience, 4
Consequences of Phenomenology (Ihde), 106
Consequences of Pragmatism (Rorty), 105–6
Crab Nebula, *120*
craft technology, 7
The Crisis of European Sciences and Transcendental Phenomenology (Husserl), 9
critical hermeneutics, 113

Data, Instruments, and Theory (Ackermann), 22
deconsruction, 5
Descartes, René, 3–4; camera obscura, 138; Husserl and, xiii
Dewey, John, 4; consciousness in, 92–99; experience, 97–98; Mitcham on, 104;

naturalization, 97–98; pragmatism and, 11, 16–17; "Psychology and Philosophic Method," 96–97; temporality, 98
Dialogue on the Two Chief World Systems (Galilei), 27–28
discovery, technologies and, 2–3
dissemination of information, 74; Greeks, 74–75
duck-rabbit, 111–12
Duhem, Pierre, 38

embodiment postphenomenology, 10–11, 75–76, 113
embodiment practices, contemporary, 75–76
empiricism, 39–40
Enlightenment humanism, 5
epistemology engines, 137–39
existential phenomenology, 124
existentialism, xii; phenomenology and, xii
Expanding Hermeneutics: Visualism in Science (Ihde), 116
experience: consciousness-centered, 4; Dewey, 93, 97–98; experimental version, 4; instrumental version, 4; learned *versus* unlearned observer, 72; naturalized, 110; postphenomenology and, 111; relativity and, 29–32; whole body, 110
Experimental Phenomenology (Ihde), 111, 128–29
eyeglasses, 14

fallibilism of science, 22
feminists, contemporary, xvi–xvii
forgetfulness of lifeworld, 55–57
Forman, Paul: *History and Technology*, 5, 7–8; science *versus* technology, 7–9
Fox-Keller, Evelyn, 21–22
freeze-frame notion of phenomenology, 105–6

Galilei, Galileo, 2–4; ahistorical, 50–51; claims about telescope, 56; *Dialogue on the Two Chief World Systems*, 27–28; history, 48–49; Husserl's, 26–27, 29–32; lifeworld and, 51–52; mathematical language of nature, 47–48; motion, 57; *The Starry Messenger*, 27; telescope, 26–29; telescope as instrumental artifact, 49–50
Galison, Peter: *How Experiments End*, 22; *Image and Logic*, 22
geometry, lifeworld, 43–44
Giere, Ronald, *Scientific Perspectivism*, 23
global environment, 78–79
global measurements, synchronic, 80
Göttingen School, 37; structural phenomenologists, 103
Greenhouse Effect, classical phenomenology and, 79–82
Gross, Paul, *Higher Superstition: The Academic Left and Its Quarrels with Science*, 5

Hacking, Ian, *Representing and Intervening*, 22
Haraway, Donna, xvi–xvii, 22, 46

Harding, Sandra, 22
Hasse, Cathrine, 127, 133
Hawking, Stephen, xviii
Heidegger, Martin: Husserl comparison, xiii; motion as seen by Galileo and Aristotle, 57
Heidegger's Technologies: Postphenomenological Perspectives (Ihde), xv–xvi, 1
hermeneutic phenomenology, 124
hermeneutics: material, 84; postphenomenological, 83–86; in technoscience, 116–22
Hickman, Larry: *John Dewey's Pragmatic Technology*, 114; *Philosophical Tools for Technological Culture*, 114
Higher Superstition: The Academic Left and Its Quarrels with Science (Norman and Paul), 5
historical lifeworld, 42
histories, 50–51
History and Technology (Forman), 5, 7–8
Holenstein, Elmar, 103–4
Hottois, Gilbert, 123
How Experiments End (Galison), 22
Hughes, Thomas, 123
The Human Condition (Arendt), xiv
humanists, 5
Humans, Nature and, 6
Husserl, Edmund: *Cartesian Meditations*, 93, 94–96, 130–32; consciousness in, 92–99; *The Crisis of European Sciences and Transcendental*

Husserl, Edmund (*continued*)
  Phenomenology, 9; Heidegger
  comparison, xiii; invention of
  phenomenology, xii; linguistic
  strategy, xiii; misinterpretation,
  terminology and, 104–5; *The
  Origin of Geometry*, 9; *The Paris
  Lectures*, 93; technologies, 9
Husserl Circle, 9–10

Ihde, Don: *Bodies in Technology*,
  116; *Chasing Technoscience:
  Matrix for Materiality*,
  105; *Consequences of
  Phenomenology*, 106;
  *Expanding Hermeneutics:
  Visualism in Science*, 116;
  Experimental Phenomenology,
  111, 128–29; *Heidegger's
  Technologies:
  Postphenomenological
  Perspectives*, xv–xvi, 1;
  *Instrumental Realism: The
  Interface Between Philosophy
  of Science and Philosophy
  of Technology*, 22, 110;
  *On Non-Foundational
  Phenomenology*, 125;
  *Postphenomenology: Essays
  in a Postmodernist Context*,
  106, 125; *Postphenomenology:
  Essays in the Postmodern
  Era*, 5; *Technics and Praxis:
  A Philosophy of Technology*,
  10; *Technology and the
  Lifeworld*, 111
Image and Logic (Galison),
  22
imaginative variations in
  phenomenology, 84–85

imaging technologies, 83–84,
  115–16; climate change and,
  120–21; computer processes,
  119–20; instrumental
  phenomenological variations,
  119; lenses, 116–18; X-ray,
  117–18
instrumental artifacts, 49–50, 72
instrumental phenomenological
  variations, 119
*Instrumental Realism: The
  Interface Between Philosophy of
  Science and Philosophy of
  Technology* (Ihde), 22, 110
instruments, 10, 24; astronomy,
  116–17; design advances, 20;
  knowledge and, 29–32; as
  mediators of meaning, 54;
  phenomenology of use, 25–26;
  philosophy of science and, 22;
  in practice *versus* as objects,
  25–26; technology and, 3–4
intake of information, 74
intentionality: *Cartesian
  Meditations* (Husserl), 130–32;
  consciousness and, 95; as
  interrelational ontology, 130
interactionism, 104
interpreters of science, 18–20
interrelational ontology of science
  and technology, 8–9;
  intentionality as, 130
intuition, 114
invariants, 85, 128
is/ought, fact/value distinction of
  pragmatism, 108

James, William, 103
*John Dewey's Pragmatic
  Technology* (Hickman), 114

Kierkegaard, Søren, xvi
Kim, Sung Dong, 125–26
Kohak, Erazim, xii
Kuhn, Thomas, 20; motion as seen by Galileo and Aristotle, 57; *Structure of Scientific Revolutions*, 19

laboratories, scientific knowledge and, 21
Laboratory Life (Latour), 6, 21
Lakatos, Imre, 18, 20
Latour, Bruno, 46; *Laboratory Life*, 6, 21; The Modern Constitution, 6–7; *We Were Never Modern*, 5, 6
Laudan, Larry, *Science and Relativism*, 22
Levitt, Norman, *Higher Superstition: The Academic Left and Its Quarrels with Science*, 5
lifeworld, 41–43, 100–2; *Crisis* and, 100–1; forgetfulness, 55–57; Galileo and, 51–52; geometry, 43–44; idealization, 44–45; *versus* science, 43–46
linguistics, xiii
Locke, John, 3; camera obscura, 138
Luckmann, Thomas, 104; *The Social Construction of Reality*, 19

Mach, Ernst, 38
magnifying glass, 67–68
Marburger, John, 129
material hermeneutics, 84
materialization, writing and, 54–55

mathematical language of nature, 47–48
mathematization of science, 12, 18, 38–39
Matisse, Henri: Blue Nude IV, *71*; cut-outs, 70
McMullin, Ernan, *The Social Dimensions of Science*, 22
Mead, George Herbert, 104
meaning, mediators, 54–55
mediation: instruments and, 54; telescope, 56–57
Merton, Robert, 19–20
metaphysics, 6
missing technologies, 13
Mitcham, Carl, *Thinking through Technology*, 104
The Modern Constitution, 6–7
modernism: Dewey, 93–94; Husserl, 93–94; science and, 5; technology and, 5
modernity: early, lenses and, 117–18; Forman and, 7–9; Latour and, 8; The Modern Constitution, 6–7; science *versus* technology, 7–9; sciences and, 6; subject-object model, 4. *See also* antimodernities, postmodernism
Moralizing Technology (Verbeek), 127
motion: Aristotle on, 57; Galileo on, 57
multistabilities, 85, 113, 128, 129–30; multistable doodle, *112*

Nachlass, 14–15, 128
natural philosophy, 123

naturalization, 110; Dewey, 97–98; nonreductive, 100–2; reductionist, 101–2
Nature: beasts and, 6–7; Humans and, 6
Natur/Geist, 69
neopragmatism, 4
nonreductive naturalism, 100–2
Nye, David, 123

observers, learned *versus* unlearned, 72
obsolescence of technologies, 15–17
Olesen, Finn, 127
On Non-Foundational Phenomenology (Ihde), 125
optical technologies, 14–15; eyeglasses, 14; timeline, 64
optics, 3
ordinary technologies, 14
The Origin of Geometry (Husserl), 9

The Paris Lectures (Husserl), 93
Peirce, Charles Sanders, 103
pens, 61–63, 70; Husserl's use, 61–63
perceptual lifeworld, 41–42
The Phenomenological Movement: A Historical Introduction (Spiegelberg), xii
phenomenology, xi–xii; contrarian epistemology of Husserl, 3–4; enriching pragmatism, 110–14; existential, 124; existentialism and, xii; freeze-frame notion, 105–6; Greenhouse Effect and, 79–82; hermeneutic, 124;

imaginative variations, 84–85; of instrument use, 25–26; misinterpretations, 104–5; origins, 3–4; positivism and, 40; structural phenomenologists, 103; subjectivity, 93; of technics, xv, 115; term, 123; transcendental, 124; of work, xiv–xv
Philosophical Tools for Technological Culture (Hickman), 114
philosophies: analytic philosophy, 108–9; natural philosophy, 123; science and, 18; of technology, 21; use-life, 17
Philosophy and the Mirror of Nature (Rorty), 105
philosophy distribution, 74–75
philosophy of science, 22; antipositivists, 40–41; Duhem, Pierre, 38; empiricism, 39–40; Husserl's, 23–25; Mach, Ernst, 38; mathematizers, 38–39; Poincaré, Jules, 38; positivists, 39–40; twentieth-century, 38–41
Philosophy of Science Association, 126–27
physical theory, science as, 38
physics, 38–39
physiological knowledge, 66–67
Pinch, Trevor, 46
Platonic tradition, pragmatism and, 107
Poincaré, Jules, 38
Popper, Karl, 18, 20
positivism, 22, 39–40; antipositivists, 40–41

positivist-antipositivist
controversies, 18
postanalytic pragmatism,
108–9, 124; *versus*
postphenomenological,
109–10
posthumanism, 5
postmodern science, 33–34
postmodernism, 5
postphenomenological
pragmatism, *versus*
postanalytic, 109–10
postphenomenology, 4, 16–17,
59; critical hermeneutics,
113; embodiment
postphenomenology, 10–11,
113; experience and, 111;
multistabilities, 85; science
praxis as hermeneutic, 83–86;
Technoscience Research Group
and, 125–26; term, 124, 125;
variational theory and, 111–13
*Postphenomenology: Essays in the
Postmodern Context* (Ihde), 5,
106, 125
poststructuralism, 5
pragmatism, 11, 16–17; anti-
essentialism, 106–7; Dewey
and, 11; enrichment through
phenomenology, 110–14; is/
ought, fact/value distinction,
108; neopragmatism, 4;
phenomenology and, 4,
103–4; Platonic tradition and,
107; postanalytic, 108–9
praxical lifeworld, 42
praxis: science, as
postphenomenological
hermeneutic, 83–86; shift
to, 100–1

praxis-originated thinking, 10, 12
psychologism, 103
psychology, consciousness and, 89
"Psychology and Philosophic
Method" (Dewey), 96–97

radio astronomy, 118
reading technologies, timeline, 64
reductionist naturalization,
101–2
relativity, unity of experiences
and, 29–32
Representing and Intervening
(Hacking), 22
res extensa, xv
Riphagen, Monique, 125–26
Rorty, Richard, 4; anti-
essentialism, 106–7;
Consequences of Pragmatism,
105–6; phenomenology
as foundational, 105–6;
*Philosophy and the Mirror
of Nature,* 105; philosophy
groupings, 105–6; postanalytic
philosophy, 124–25

Schaffer, Simon, 46
science: applied, 7; Big Science,
35–36; fallibilism, 22; *versus*
lifeworld, 43–44; as
mathematization, 12;
modernity and, 6;
philosophies and, 18; as
physical theory, 38;
postmodern, 33–34; sociology
of, 19; *versus* technology, 7–8;
term, 123; unified, 39–40.
See also technoscience
Science and Relativism (Laudan),
22

Perspectives in
Continental Philosophy
John D. Caputo, series editor

Jean-Luc Marion, *In Excess: Studies of Saturated Phenomena.* Translated by Robyn Horner and Vincent Berraud.

Phillip Goodchild, *Rethinking Philosophy of Religion: Approaches from Continental Philosophy.*

William J. Richardson, S.J., *Heidegger: Through Phenomenology to Thought.*

Jeffrey Andrew Barash, *Martin Heidegger and the Problem of Historical Meaning.*

Jean-Louis Chrétien, *Hand to Hand: Listening to the Work of Art.* Translated by Stephen E. Lewis.

Jean-Louis Chrétien, *The Call and the Response.* Translated with an introduction by Anne Davenport.

D. C. Schindler, *Han Urs von Balthasar and the Dramatic Structure of Truth: A Philosophical Investigation.*

Julian Wolfreys, ed., *Thinking Difference: Critics in Conversation.*

Allen Scult, *Being Jewish/Reading Heidegger: An Ontological Encounter.*

Richard Kearney, *Debates in Continental Philosophy: Conversations with Contemporary Thinkers.*

Jennifer Anna Gosetti-Ferencei, *Heidegger, Hölderlin, and the Subject of Poetic Language: Toward a New Poetics of Dasein.*

Jolita Pons, *Stealing a Gift: Kierkegaard's Pseudonyms and the Bible.*

Jean-Yves Lacoste, *Experience and the Absolute: Disputed Questions on the Humanity of Man.* Translated by Mark Raftery-Skehan.

Charles P. Bigger, *Between* Chora *and the Good: Metaphor's Metaphysical Neighborhood.*

Dominique Janicaud, *Phenomenology "Wide Open": After the French Debate.* Translated by Charles N. Cabral.

Ian Leask and Eoin Cassidy, eds., *Givenness and God: Questions of Jean-Luc Marion.*

Jacques Derrida, *Sovereignties in Question: The Poetics of Paul Celan.* Edited by Thomas Dutoit and Outi Pasanen.

William Desmond, *Is There a Sabbath for Thought? Between Religion and Philosophy.*

Bruce Ellis Benson and Norman Wirzba, eds., *The Phenomenology of Prayer.*

S. Clark Buckner and Matthew Statler, eds., *Styles of Piety: Practicing Philosophy after the Death of God.*

Kevin Hart and Barbara Wall, eds., *The Experience of God: A Postmodern Response.*

John Panteleimon Manoussakis, *After God: Richard Kearney and the Religious Turn in Continental Philosophy.*

John Martis, *Philippe Lacoue-Labarthe: Representation and the Loss of the Subject.*

Jean-Luc Nancy, *The Ground of the Image.*

Edith Wyschogrod, *Crossover Queries: Dwelling with Negatives, Embodying Philosophy's Others.*

Gerald Bruns, *On the Anarchy of Poetry and Philosophy: A Guide for the Unruly.*

Brian Treanor, *Aspects of Alterity: Levinas, Marcel, and the Contemporary Debate.*

Simon Morgan Wortham, *Counter-Institutions: Jacques Derrida and the Question of the University.*

Leonard Lawlor, *The Implications of Immanence: Toward a New Concept of Life.*

Clayton Crockett, *Interstices of the Sublime: Theology and Psychoanalytic Theory.*

Bettina Bergo, Joseph Cohen, and Raphael Zagury-Orly, eds., *Judeities: Questions for Jacques Derrida.* Translated by Bettina Bergo and Michael B. Smith.

Jean-Luc Marion, *On the Ego and on God: Further Cartesian Questions.* Translated by Christina M. Gschwandtner.

Jean-Luc Nancy, *Philosophical Chronicles.* Translated by Franson Manjali.

Jean-Luc Nancy, *Dis-Enclosure: The Deconstruction of Christianity.* Translated by Bettina Bergo, Gabriel Malenfant, and Michael B. Smith.

Andrea Hurst, *Derrida Vis-à-vis Lacan: Interweaving Deconstruction and Psychoanalysis.*

Jean-Luc Nancy, *Noli me tangere: On the Raising of the Body.* Translated by Sarah Clift, Pascale-Anne Brault, and Michael Naas.

Jacques Derrida, *The Animal That Therefore I Am*. Edited by Marie-Louise Mallet, translated by David Wills.

Jean-Luc Marion, *The Visible and the Revealed*. Translated by Christina M. Gschwandtner and others.

Michel Henry, *Material Phenomenology*. Translated by Scott Davidson.

Jean-Luc Nancy, *Corpus*. Translated by Richard A. Rand.

Joshua Kates, *Fielding Derrida*.

Michael Naas, *Derrida From Now On*.

Shannon Sullivan and Dennis J. Schmidt, eds., *Difficulties of Ethical Life*.

Catherine Malabou, *What Should We Do with Our Brain?* Translated by Sebastian Rand, Introduction by Marc Jeannerod.

Claude Romano, *Event and World*. Translated by Shane Mackinlay.

Vanessa Lemm, *Nietzsche's Animal Philosophy: Culture, Politics, and the Animality of the Human Being*.

B. Keith Putt, ed., *Gazing Through a Prism Darkly: Reflections on Merold Westphal's Hermeneutical Epistemology*.

Eric Boynton and Martin Kavka, eds., *Saintly Influence: Edith Wyschogrod and the Possibilities of Philosophy of Religion*.

Shane Mackinlay, *Interpreting Excess: Jean-Luc Marion, Saturated Phenomena, and Hermeneutics*.

Kevin Hart and Michael A. Signer, eds., *The Exorbitant: Emmanuel Levinas Between Jews and Christians*.

Bruce Ellis Benson and Norman Wirzba, eds., *Words of Life: New Theological Turns in French Phenomenology*.

William Robert, *Trials: Of Antigone and Jesus*.

Brian Treanor and Henry Isaac Venema, eds., *A Passion for the Possible: Thinking with Paul Ricoeur*.

Kas Saghafi, *Apparitions—Of Derrida's Other*.

Nick Mansfield, *The God Who Deconstructs Himself: Sovereignty and Subjectivity Between Freud, Bataille, and Derrida*.

Don Ihde, *Heidegger's Technologies: Postphenomenological Perspectives*.

Suzi Adams, *Castoriadis's Ontology: Being and Creation*.

Richard Kearney and Kascha Semonovitch, eds., *Phenomenologies of the Stranger: Between Hostility and Hospitality*.

Michael Naas, *Miracle and Machine: Jacques Derrida and the Two Sources of Religion, Science, and the Media*.

Alena Alexandrova, Ignaas Devisch, Laurens ten Kate, and Aukje van Rooden, *Re-treating Religion: Deconstructing Christianity with Jean-Luc Nancy*. Preamble by Jean-Luc Nancy.

Emmanuel Falque, *The Metamorphosis of Finitude: An Essay on Birth and Resurrection*. Translated by George Hughes.

Scott M. Campbell, *The Early Heidegger's Philosophy of Life: Facticity, Being, and Language*.

Françoise Dastur, *How Are We to Confront Death? An Introduction to Philosophy*. Translated by Robert Vallier. Foreword by David Farrell Krell.

Christina M. Gschwandtner, *Postmodern Apologetics? Arguments for God in Contemporary Philosophy*.

Ben Morgan, *On Becoming God: Late Medieval Mysticism and the Modern Western Self*.

Neal DeRoo, *Futurity in Phenomenology: Promise and Method in Husserl, Levinas, and Derrida*.

Sarah LaChance Adams and Caroline R. Lundquist, eds., *Coming to Life: Philosophies of Pregnancy, Childbirth, and Mothering*.

Thomas Claviez, ed., *The Conditions of Hospitality: Ethics, Politics, and Aesthetics on the Threshold of the Possible*.

Roland Faber and Jeremy Fackenthal, eds., *Theopoetic Folds: Philosophizing Multifariousness*.

Jean-Luc Marion, *The Essential Writings*. Edited by Kevin Hart.

Adam S. Miller, *Speculative Grace: Bruno Latour and Object-Oriented Theology*. Foreword by Levi R. Bryant.

Jean-Luc Nancy, *Corpus II: Writings on Sexuality*.

David Nowell Smith, *Sounding/Silence: Martin Heidegger at the Limits of Poetics*.

Gregory C. Stallings, Manuel Asensi, and Carl Good, eds., *Material Spirit: Religion and Literature Intranscendent*.

Claude Romano, *Event and Time*. Translated by Stephen
E. Lewis.

Frank Chouraqui, *Ambiguity and the Absolute: Nietzsche and
Merleau-Ponty on the Question of Truth*.

Noëlle Vahanian, *The Rebellious No: Variations on a Secular
Theology of Language*.

Michael Naas, *The End of the World and Other Teachable
Moments: Jacques Derrida's Final Seminar*.

Jean-Louis Chrétien, *Under the Gaze of the Bible*. Translated by
John Marson Dunaway.

Edward Baring and Peter E. Gordon, eds., *The Trace of God:
Derrida and Religion*.

Vanessa Lemm, ed., *Nietzsche and the Becoming of Life*.

Aaron T. Looney, *Vladimir Jankélévitch: The Time of Forgiveness*.

Richard Kearney and Brian Treanor, eds., *Carnal Hermeneutics*.

Tarek R. Dika and W. Chris Hackett, *Quiet Powers of the
Possible: Interviews in Contemporary French Phenomenology*.
Foreword by Richard Kearney.

Jeremy Biles and Kent L. Brintnall, eds., *Georges Bataille and the
Study of Religion*.

William S. Allen, *Aesthetics of Negativity: Blanchot, Adorno, and
Autonomy*.

Don Ihde, *Husserl's Missing Technologies*.

Colby Dickinson and Stéphane Symons (eds.), *Walter Benjamin
and Theology*.

Emmanuel Falque, *Crossing the Rubicon: The Borderlands of
Philosophy and Theology*. Translated by Reuben Shank.
Introduction by Matthew Farley.